John M. Tyler

The Whence and the Whither of Man

A brief history of his origin and development through conformity to environment.

Being the Morse lectures of 1895

John M. Tyler

The Whence and the Whither of Man
A brief history of his origin and development through conformity to environment. Being the Morse lectures of 1895

ISBN/EAN: 9783337212827

Printed in Europe, USA, Canada, Australia, Japan

Cover: Foto ©ninafisch / pixelio.de

More available books at **www.hansebooks.com**

THE WHENCE AND THE WHITHER OF MAN

THE WHENCE AND THE WHITHER OF MAN

A BRIEF HISTORY OF HIS ORIGIN AND DEVELOPMENT THROUGH CONFORMITY TO ENVIRONMENT

Being the Morse Lectures of 1895

BY

JOHN M. TYLER

PROFESSOR OF BIOLOGY, AMHERST COLLEGE

NEW YORK

CHARLES SCRIBNER'S SONS

1899

TROW DIRECTORY
PRINTING AND BOOKBINDING COMPANY
NEW YORK

TABLE OF CONTENTS

CHAPTER IV

The advance of vertebrates from fish through amphibia and reptiles to mammals.—The development of skeleton, appendages, circulatory and respiratory systems, and brain.—Mammals: The oviparous monotremata. —Marsupials.—Placental mammals.—Development of the placenta.—Primates.—Arboreal life and the development of the hand.—Comparison of man with the highest apes.—Recapitulation of the history of man's origin and development.—The sequence of dominant functions.

CHAPTER V

Mode of investigation.—Intellect.—Sense-perceptions.—Association.— Inference and understanding.—Rational intelligence.—Modes of mental or nervous action.—Reflex action, unconscious and comparatively mechanical. —Instinctive action: The actor is conscious, but guided by heredity.—Intelligent action.—The actor is conscious, guided by intelligence resulting from experience or observation.—The will stimulated by motives.—Appetites.—Fear and other prudential considerations.—Care for young and love of mates.—The dawn of unselfishness.—Motives furnished by the rational intelligence: Truth, right, duty.—Recapitulation: The will, stimulated by ever higher motives, is finally to be dominated by unselfishness and love of truth and righteousness.—These rouse the only inappeasable hunger, and are capable of indefinite development.—Strength of these motives.— Their complete dominance the goal of human development.

CHAPTER VI

The reversal of the sequence of functions leads to extermination, degeneration, or, rarely, to stagnation.—Natural selection becomes more unsparing as we go higher.—Extinction.—Severity of the struggle for life.— Environment one.—But lower animals come into vital relation with but a small part of it.—It consists of a myriad of forces, which, as acting on a given form, may be considered as one grand resultant.—Environment is thus a power making at first for digestion and reproduction, then for muscular strength and activity, then for shrewdness, finally for unselfishness and righteousness.—An ultimate "power, not ourselves, making for righteousness," a personality.—Our knowledge of this personality may be valid, even though very incomplete.—Religion.—Conformity to the spiritual in or behind environment is likeness to God.—The conservative tendency in evolution.

Human environment.—The development of the family as the school of man's training.—The family as the school of unselfishness and obedience. —The family as the basis of social life.—Society as an aid to conformity to environment by increasing intelligence and training conscience.—Mental and moral heredity.—Personal magnetism.—Man's search for a king.— The essence of Christianity.—Conformity to environment gives future supremacy, but often at the cost of present hardship.—Conformity as obedience to the laws of our being.—Environment best understood through the study of the human mind.—Productiveness and prospectiveness of vital capital.—Faith.

Composed of atoms and molecules, hence subject to chemical and physical laws.—As a living being.—As an animal.—As a vertebrate.—As a mammal.—As a social being.—As a personal and moral being.—The conflict between the higher and the lower in man.—As a religious being.—As hero.—He has not yet attained.—Future man.—He will utilize all his powers, duly subordinating the lower to the higher.—The triumph of the common people.

Subject of the Bible.—*Man:* Body, intellect, heart.—*God:* Law, sin, and penalty.—God manifested in Christ.—Salvation, the divine life permeating man.—Faith.—Prayer.—Hope.—The Church.—The battle.—The victory.—The crown.

CHAPTER X

INTRODUCTION

In the year 1865 Professor Samuel Finley Breese Morse, to whom the world is indebted for the application of the principles of electro-magnetism to telegraphy, gave the sum of ten thousand dollars to Union Theological Seminary to found a lectureship in memory of his father, the Rev. Jedediah Morse, D.D., theologian, geographer, and gazetteer. The subject of the lectures was to have to do with " The relations of the Bible to any of the sciences." The ten chapters of this book correspond to ten lectures, eight of which were delivered as Morse Lectures at Union Theological Seminary during the early spring of 1895. The first nine chapters appear in form and substance as they were given in the lectures, except that Chapters VI. and VII. were condensed in one lecture. Chapter X. is new, and I have not hesitated to add a few paragraphs wherever the argument seemed especially to demand further evidence or illustration.

One of my friends, reading the title of these lectures, said : " Of man's origin you know nothing, of his future you know less." I fear that many share his opinion, although they might not express it so emphatically.

It would seem, therefore, to be in order to show that science is now competent to deal with this question ; not that she can give a final and conclusive answer, but that we can reach results which are probably in the main correct.) We may grant very cheerfully that we can attain no demonstration ; the most that we can claim for our results will be a high degree of probability. If our conclusions are very probably correct, we shall do well to act according to them; for all our actions in life are suited to meet the emergencies of a probable but uncertain course of events.

We take for granted the probable truth of the theory of evolution as stated by Mr. Darwin, and that it applies to man as really as to any lower animal. At the same time it concerns our argument but little whether natural selection is " omnipotent " or of only secondary importance in evolution, as long as it is a real factor, or which theory of heredity or variation is the more probable.

If man has been evolved from simple living substance protoplasm, by a process of evolution, it will some day be possible to write a history of that process. But have we yet sufficient knowledge to justify such an attempt?

Before the history of any period can be written its events must have been accurately chronicled. Biological history can be written only when the successive stages of development and the attainments of each stage have been clearly perceived. In other words, the first prerequisite would seem to be a genealogical * tree of the animal kingdom. The means of tracing this genealogical tree are given in the first chapter, and the results in the second, third, and fourth chapters of this book.

Now, for some of the ancestral stages of man's development a very high degree of probability can be

* See Phylogenetic Chart, p. 310.

claimed. One of man's earliest ancestors was almost certainly a unicellular animal.) A little later he very probably passed through a gastræa stage. He traversed fish, amphibian, and reptilian grades. The oviparous monotreme and the marsupial almost certainly represent lower mammalian ancestral stages. But what kind of fish, what species of amphibian, what form of reptiles most closely resembles the old ancestor? How did each of these ancestors look? I do not know. It looks as if our ancestral tree were entirely uncertain and we were left without any foundation for history or argument.

But the history of the development of anatomical details, however important and desirable, is not the only history which can be written, nor is it essential. It would be interesting to know the size of brain, girth of chest, average stature, and the features of the ancient Greeks and Romans. But this is not the most important part of their history, nor is it essential. The great question is, What did they contribute to human progress?

Even if we cannot accurately portray the anatomical details of a single ancestral stage, can we perhaps discover what function governed its life and was the aim of its existence? Did it live to eat, or to move, or to think? If we cannot tell exactly how it looked, can we tell what it lived for and what it contributed to the evolution of man?

Now, the sequence of dominant functions or aims in life can be traced with far more ease and safety, not to say certainty, than one of anatomical details. The latter characterize small groups, genera, families, or classes; while the dominant function characterizes all animals of a given grade, even those which through degeneration have reverted to this grade.

Even if I cannot trace the exact path which leads to the mountain-top, I may almost with certainty affirm that it leads from meadow and pasture through forest to bare rock, and thence over snow and ice to the summit; for each of these forms a zone encircling the mountain. Very similarly I find that, whatever genealogical tree I adopt, one sequence in the dominance of functions characterizes them all; digestion is dominant before locomotion and locomotion before thought.

And it is hardly less than a physiological necessity that it should be so. The plant can and does exist, living almost purely for digestion and reproduction, and the same is true of the lowest and most primitive animals. A muscular system cannot develop and do its work until some sort of a digestive system has arisen to furnish nutriment, any more than a steam-engine can run without fuel. And a brain is of no use until muscle and sense-organs have appeared.

This sequence of dominant functions,* of physiological dynasties, would seem therefore to be a fact. And our series of forms described in the second, third, and fourth chapters is merely a concrete illustration showing how this sequence may have been evolved. The substitution of other terms in the anatomical series there described—amœba, volvox, etc.—would not affect this result. By a change in the form of our history we have eliminated to a large extent the sources of uncertainty and error. And the dominant function of a group throws no little light on the details of its anatomy.

* See condensed Chart of Development, etc., p. 309.

If we can be satisfied that ever higher functions have risen to dominance in the successive stages of animal and human development, if we can further be convinced that the sequence is irreversible, we shall be convinced that future man will be more and more completely controlled by the very highest powers or aims to which this sequence points. Otherwise we must disbelieve the continuity of history. But the germs of the future are always concealed in the history of the present. Hence—pardon the reiteration—if we can once trace this sequence of dominant functions, whose evolution has filled past ages, we can safely foretell something at least of man's future development.

The argument and method is therefore purely historical. Here and there we will try to find why and how things had to be so. But all such digressions are of small account compared with the fact that things were or are thus and so. And a mistaken explanation will not invalidate the facts of history.

The subject of our history is the development, not of a single human race nor of the movements of a century, but the development of animal life through ages. And even if our attempts to decipher a few pages here and there in the volumes of this vast biological history are not as successful as we could hope, we must not allow ourselves to be discouraged from future efforts. Even if our translation is here and there at fault, we must never forget the existence of the history. Some of the worst errors of biologists are due to their having forgotten that in the lower stages the germs of the higher must be present, even though invisible to any microscope. Our study of the worm is inadequate and likely to mislead us, unless we remember that a worm was the ancestor of man. And a biologist who can tell us nothing about man is neglecting his fairest field.

Conversely history and social science will rest on a firmer basis when their students recognize that many human laws and institutions are heirlooms, the attainments, or direct results of attainments, of animals far below man. We are just beginning to recognize that the study of zoölogy is an essential prerequisite to, and firm foundation for, that of history, social science, philosophy, and theology, just as really as for medicine. An adequate knowledge of any history demands more than the study of its last page. The zoölogist has been remiss in not claiming his birthright, and in this respect has sadly failed to follow the path pointed out by Mr. Darwin.

For palæontology, zoölogy, history, social and political science, and philosophy are really only parts of one great science, of biology in the widest sense, in distinction from the narrower sense in which it is now used to include zoölogy and botany. They form an organic unity in which no one part can be adequately understood without reference to the others. You know nothing of even a constellation, if you have studied only one of its stars. Much less can the study of a single organ or function give an adequate idea of the human body.

Only when we have attained a biological history can we have any satisfactory conception of environment. As we look about us in the world, environment often seems to us to be a chaos of forces aiding or destroying good and bad, fit and unfit, alike.

But our history of animal and human progress shows us successive stages, each a little higher than the preceding, and surviving, for a time at least, because more completely conformed to environment. If this be true, and it must be true unless our theory of evolution be false, higher forms are more completely conformed to their environment than lower; and man has attained the most complete conformity of all. Our biological history is therefore a record of the results of successive efforts, each attaining a little more complete conformity than the preceding. From such a history we ought to be able to draw certain valid deductions concerning the general character and laws of our environment, to discover the direction in which its forces are urging us, and how man can more completely conform to it.

If man is a product of evolution, his mental and moral, just as really as his physical, development must be the result of such a conformity. The study of environment from this standpoint should throw some light on the validity of our moral and religious creeds and theories. It would seem, therefore, not only justifiable, but imperative to attempt such a study.

Our argument is not directly concerned with modern theories of heredity, or variation, or with the "omnipotence" or secondary importance of natural selection. And yet Nägeli, and especially Weismann, have had so marked an influence on modern thought that we cannot afford to neglect their theories. We will briefly notice these in the closing chapter.

CHAPTER I

THE story of a human life can be told in very few words. A youth of golden dreams and visions ; a few years of struggle or of neglected opportunities; then retrospect and the end.

" We come like water, and like wind we go."

But how few of the visions are realized. Faust sums up the whole of life in the twice-repeated word *versagen*, renounce, and history tells a similar story. Terah died in Haran ; Abraham obtained but a grave in the land promised him and his children; Jacob, cheated in marriage, bitterly disappointed in his children, died in exile, leaving his descendants to become slaves in the land of Egypt; and Moses, their heroic deliverer, died in the mountains of Moab in sight of the land which he was forbidden to enter. You may answer that it is no injury that the promise is too large, the vision too grand, to be fulfilled in the span of a single life, but must become the heritage of a race. But what has been the history of Abraham's descendants? A death-grapple for existence, captivity, and dispersion. Their national existence has long been lost.

Was there ever a nation of grander promise than Greece or Rome ? But Greece died of premature old age, and Rome of rottenness begotten of sin. But each of them, you will say, left a priceless heritage to the immortal race. But if Greece and Rome and a host of older nations, of which History has often forgotten the very name, have failed and died, can anything but ultimate failure await the race? Is human history to prove a story told by an idiot, or does it "signify" something? Is the great march of humanity, which Carlyle so vividly depicts, "from the inane to the inane, or from God to God?"

This is the sphinx question put to every thinking man, and on his answer hangs his life. For according to that answer, he will either flinch and turn back, or expend every drop of blood and grain of power in urging on the march.

To this question the Bible gives a clear and emphatic answer. "God created man in his own image," and then, as if men might refuse to believe so astounding a statement, it is repeated, "in the image of God created he him." When, and by what mode or process, man was created we are not told. His origin is condensed almost into a line, his present and future occupy all the rest of the book. Whence we came is important only in so far as it teaches us humility and yet assures us that we may be Godlike because we are His handiwork and children, "heirs of God and joint heirs with Christ of a heavenly inheritance."

Now has Science any answer to this vital question? Perhaps. But this much is certain; it can foretell the future only from the past. Its answer to the question *whither* must be an inference from its knowledge as to *whence* we have come. The Bible looks mainly at the present and future; Science must at least begin with the study of the past. The deciphering of man's past history is the great aim of Biology, and ultimately of all Science. For the question of Man's past is only a part of a greater question, the origin of all living species.

We may say broadly that concerning the origin of species two theories, and only two, seem possible. The first theory is that every species is the result of an act of immediate creation. And every true species, however slightly it may differ from its nearest relative, represents such a creative act, and once created is practically unchangeable. This is the theory of immutability of species. According to the second theory all higher, probably all present existing, species are only mediately the result of a creative act. The first living germ, whenever and however created, was infused with power to give birth to higher species. Of these and their descendants some would continue to advance, others would degenerate. Each theory demands equally for its ultimate explanation a creative act; the second as much as, if not more than, the first. According to the first theory the creative power has been distributed over a series of acts, according to the second theory it has been concentrated in one primal creation. The second is the theory of the mutability of species, or, in general, of evolution, but not necessarily of Darwinism alone.

The first theory is considered by many the more attractive and hopeful. Now a theory need not be attractive, nor at first sight appear hopeful, provided only it is true. But let me call your attention to certain conclusions which, as it appears to me, are necessarily involved in it. Its central thought is the practical immutability of species. Each one of these lives its little span of time, for species are usually comparatively short-lived, grows possibly a very little better or worse, and dies. Its progress has added nothing to the total of life; its degeneration harmed no one, hardly even itself; it was doomed from the start. Progress there has been, in a sense. The Creator has placed ever higher forms on the globe. But all the progress lies in the gaps and distances between successive forms, not in any advance made, or victory won, by the species or individual. The most "aspiring ape," if ever there was such a being, remains but an ape. He must comfort himself with the thought that, while he and his descendants can never gain an inch, the gap between himself and the next higher form shall be far greater than that between himself and the lowest monkey.

And if this has been the history of thousands of other species, why should it not be true of man also? Who can wonder that many who accept this theory doubt whether the world is growing any better, or whether even man will ever be higher and better than he now is? Would it not be contrary to the whole course of past history, if you can properly call such a record a history, if he could advance at all? Now I have no wish to misrepresent this or any honestly accepted theory, but it appears to me essentially hopeless, a record not of the progress of life on the globe, but of a succession of stagnations, of deaths. I can never understand why some very good and intelligent people still think that the theory of the immediate creation of each species does more honor to the Creator and his creation than the theory of evolution. Evo-

lution is a process, not a force. The power of the
Creator is equally demanded in both cases; only it is
differently distributed. And evolution is the very
highest proof of the wisdom and skill of the Creator.
It elevates our views of the living beings, must it not
give a higher conception of Him who formed them?

The plant in its first stages shows no trace of flowers,
but of leaves only. Later a branch or twig, similar in
structure to all the rest, shortens. The cells and tissues which in other twigs turn into green leaves here
become the petals and other organs of the rose or
violet. Let us suppose for a moment that every rose
and violet required a special act of immediate creation,
would the springtime be as wonderful as now? Would
the rose or violet be any more beautiful, or are they
any less flowers because developed out of that which
might have remained a common branch? The plant
at least is glorified by the power to give rise to such
beauty. And is not the creation of the seed of a violet or rose something infinitely grander than the deck-
ing of a flowerless plant with newly created roses?
The attainment of the highest and most diversified
beauty and utility with the fewest and simplest means
is always the sign of what we call in man "creative"
genius. Is not the same true of God? I think you
all feel the force of the argument here.

There were at one time no flowering plants. The
time came at last for their appearance. Which is the
higher, grander mode of producing them, immediate
creation of every flowering species, or development of
the flower out of the green leaves of some old club
moss or similar form? The latter seems to me at
least by far the higher mode. And to have created a

ground-pine which could give rise to a rose seems far more difficult and greater than to have created both separately. It requires more genius, so to speak. It gives us a far higher opinion of the ground-pine; does it disgrace the rose? We can look dispassionately at plants. The rose is still and always a rose, and the oak an oak, whatever its origin. And I believe that we shall all readily admit that evolution is here a theory which does the highest honor to the wisdom and power of the Creator. What if the animal kingdom is continually blossoming in ever higher forms? Does not the same reasoning hold true, only with added force? I firmly believe that we should all unhesitatingly answer, yes, could we but be assured that all men would everywhere and always believe that we, men, were the results of an immediate creative act.

But why do we so strenuously object to the application to ourselves of the theory of evolution? One or two reasons are easily seen. We have all of us a great deal of innate snobbery, we would rather have been born great than to have won greatness by the most heroic struggle. But is man any less a man for having arisen from something lower, and being in a fair way to become something higher? Certainly not, unless I am less a man for having once been a baby. It is only when I am unusually cross and irritable that I object to being reminded of my infancy. But a young child does not like to be reminded of it. He is afraid that some one will take him for a baby still. And the snob is always desperately afraid that some one will fail to notice what a high-born gentleman he is.

Now man can relapse into something lower than a

brute ; the only genuine brute is a degenerate man. And we all recognize the strength of tendencies urging us downward. Is not this the often unrecognized kern of our eagerness for some mark or stamp that shall prove to all that we are no apes, but men ? It is not the pure gold that needs the " guinea stamp." If we are men, and as we become men, we shall cease to fear the theory of evolution. Now this is not the only, or per-haps the greatest, objection which men feel or speak against the theory. But I must believe that it has more weight with us than we are willing to admit.

But some say that the theory of immediate creation and immutability of species is the more natural and has always been accepted, while the theory of evolu-tion is new and very likely to be as short-lived as many another theory which has for a time fascinated men only to be forgotten or ridiculed.

But the idea of evolution is as old as Hindu philoso-phy. The old Ionic natural philosophers were all evolutionists. So Aristophanes, quoting from these or Hesiod concerning the origin of things, says : " Chaos was and Night, and Erebus black, and wide Tartarus. No earth, nor air nor sky was yet ; when, in the vast bosom of Erebus (or chaotic darkness) winged Night brought forth first of all the egg, from which in after revolving periods sprang Eros (Love) the much desired, glittering with golden wings ; and Eros again, in union with Chaos, produced the brood of the human race." Here the formative process is a birth, not a creation ; it is evolution pure and simple. " According to the ancient view," says Professor Lewis, " the present world was a growth ; it was born, it came from some-thing antecedent, not merely as a cause but as its seed, embryo or principium. Plato's world was a 'zoon,' a living thing, a natural production."

Furthermore, to the ancient writers of the Bible the idea of origin by birth from some antecedent form—and this is the essential idea of evolution—was perfectly natural. They speak of the "generations of the heavens and the earth" as of the "generations" of the patriarchs. The first book of the Bible is still called Genesis, the book of births. The writer of the ninetieth Psalm says, "Before the mountains were born, or ever thou hadst brought to birth the earth and the world." And what satisfactory meaning can you give to the words, "Let the earth bring forth," and "the earth brought forth," in immediate proximity to the words, "and God made," unless while the ultimate source was God's creative power, the immediate process of formation was one of evolution.

The Bible is big and broad enough to include both ideas, the human mind is prone to overestimate the one or the other. Traces, at least, of a similar mode of thought persisted by the Greek Fathers of the Church, and disappeared, if ever, with the predominance of Latin theology. To the oriental the idea of evolution is natural. The earth is to him no inert, resistant clod ; she brings forth of herself.

But our ancestors lived on a barren soil beneath a forbidding sky. They were frozen in winter and parched in summer. Nature was to them no kind foster-mother, but a cruel stepmother, training them by stern discipline to battle with her and the world. They peopled the earth with gnomes and cobolds and giants, and their nymphs were the Valkyre. Their God was Thor, of the thunderbolt and hammer, and who

yet lived in continual dread of the hostile powers of
Nature. A Norse prophet or prophetess standing beside Elijah at Horeb would have bowed down before
the earthquake or the fire; the oriental waited for the
"still small voice." And we are heirs to a Latin
theology grafted on to the Thor-worship of our pagan
ancestors. The idea of a Nature producing benefi-
cently and kindly at the word of a loving God is
foreign to all our inherited modes of thought. And
our views of the heart of Nature are about as correct
as those of our ancestors were of God. A little more
of oriental tendencies of thought would harm neither
our theology nor our life.

What, then, is the biblical idea of Nature? God
speaks to the earth, in the first chapter of Genesis, and
the earth responds by "giving birth" to mountains and
living beings. It is evidently no mere lifeless, inert
clod, but pulsating with life and responsive to the divine commands. While yet a chaos it had been
brooded over by the Divine Spirit. It is like the great
"wheels within wheels," with rings full of eyes round
about, which Ezekiel saw in his vision by the river
Chebar. "When the living creatures went, the wheels
went by them; and when the living creatures were lifted
up from the earth, the wheels were lifted up. Whither-
soever the spirit was to go, they went, thither was their
spirit to go; and the wheels were lifted up over against
them: for the spirit of the living creatures (or of life)
was in the wheels." And above the living creatures
was the firmament and the throne of God. So Nature
may be material, but it is material interpenetrated by
the divine; if you call it a fabric, the woof may be
material but the warp is God. This view contains all

the truth of materialism and pantheism, and vastly
more than they, and it avoids their errors and omis-
sions.

To the old metaphysical hypothesis of evolution
Mr. Darwin gave a scientific basis. It had always
been admitted that species were capable of slight variation and that this divergence might become hereditary and thus perhaps give rise to a variety of the
parent species. But it was denied that the variation
could go on increasing indefinitely, it seemed soon to
reach a limit and stop. Early in the present century
Lamarck had attempted to prove that by the use and
disuse of organs through a series of generations a great
divergence might arise resulting in new species. But
the theory was crude, capable at best of but limited
application, and fell before the arguments and authority
of Cuvier. The times were not ripe for such a theory.
Some fifty years later, Mr. Darwin called attention to
the struggle for existence as a means of aggregating
these slight modifications in a divergence sufficient to
produce new species, genera, or families. His argument may be very briefly stated as follows :

1. There is in Nature a law of heredity ; like begets
like.

2. The offspring is never exactly like the parent;
and the members of the second generation differ more
or less from one another. This is especially noticeable
in domesticated plants and animals, but no less true of
wild forms. If the parent is not exactly like the
other members of the species, some of its descendants
will inherit its peculiarities enhanced, others diminished.

3. Every species tends to increase in geometrical

progression. But most species actually increase in number very slowly, if at all. Now and then some insect or weed escapes from its enemies, comes under favorable food conditions, and multiplies with such rapidity that it threatens to ravage the country. But as it multiplies it furnishes an abundance of food for the enemies which devour it, or of food and place for the parasites in and upon it; and they increase with at least equal rapidity. Hence while the vanguard increases prodigiously in numbers, because it has outrun these enemies, the rear is continually slaughtered. And thus these plagues seem in successive generations to march across the continent.

And yet even they give but a faint idea of the reproductive powers of plants and animals. The female fish produces often many thousands, sometimes hundreds of thousands of eggs. Insects generally from a hundred to a thousand. Even birds, slowly as they increase, produce in a lifetime probably at least from twelve to twenty eggs. Now let us suppose that all these eggs developed, and all the birds lived out their normal period of life, and reproduced at the same rate. After not many centuries there would not be standing room on the globe for the descendants of a single pair.

Again, of the one hundred eggs of an insect let us suppose that only sixty develop into the first larval, caterpillar, stage. Of these sixty, the number of members of the species remaining constant, only two will survive. The other fifty-eight die—of starvation, parasites, or other enemies, or from inclement weather. Now which two of all shall survive? Those naturally best able to escape their enemies or to resist unfavorable influences; in a word, those best suited to their conditions, or, to use Mr. Darwin's words, " conformed to their environment."

Now if any individual has varied so as to possess some peculiarity which enables it even in slight degree to better escape its enemies or to resist unfavorable conditions, those of its descendants who inherit most markedly this peculiar quality or variation will be the most likely to escape, those without it to perish. If a form varies unfavorably, becomes for instance more conspicuous to its enemies, it will almost certainly perish. Thus favorable variations tend to increase and become more marked from generation to generation.

Now it has always been known that breeders could produce a race of markedly peculiar form or characteristics by selecting the individuals possessing this quality in the highest degree and breeding only from these. The breeder depends upon heredity, variation, and his selection of the individuals from which to breed. Similarly in nature new species have arisen through heredity, variation, and a selection according to the laws of nature of those varying in conformity with their environment. And this Mr. Darwin called natural, in contrast with the breeder's artificial, " selection," arising from the "struggle for existence," and resulting in what Mr. Spencer has called the " survival of the fittest."

Let us take a single illustration. Many of the species of beetles on oceanic islands have very rudimentary wings, or none at all, and yet their nearest relatives are winged forms on some neighboring continent. Mr. Darwin would explain the origin of these evidently distinct wingless species as follows: They are descended from winged ancestors blown or otherwise transported thither from the neighboring continent. But beetles are slow and clumsy fliers, and on these wind-swept islands those which flew most would be blown out to sea and drowned. Those which flew the least, and these would include the individuals with more poorly developed wings, would survive. There would thus ·be a survival in every generation of a larger proportion of those having the poorest wings, and destruction of those whose wings were strong, or whose habits most active. We have here a natural selection which must in time produce a species with rudimentary or aborted wings, just as surely as a human breeder, by artificial selection can produce such an animal as a pug or a poodle. These, like sin, are a human device; nature should not be held responsible for them.

But you may urge that the variation which would take place in a single generation would be, as a rule, too slight to be of any practical value to the animal, and could not be fostered by natural selection until greatly enhanced by some other means. Let us think a moment. If ten ordinary men run in a foot-race, the two foremost may lead by several feet. But if the number of runners be continually increased the finish will be ever closer until finally but an atom more wind or muscle or pluck would make all the difference between winning and losing the prize.

Similarly the million or more young of any species of insect in a given area may be said to run a race of which the prize is life, and the losing of which means literally death. The competition is inconceivably severe. How indefinitely slight will be the difference between the poorest of the 2,000 or 20,000 survivors and the best of the more than 900,000 which perish. The very slightest favorable variation may make all the difference between life and sure death. And yet these indefinitely slight variations continued and aggregated through ages would foot up an immense total divergence. The chalk cliffs of England have been built up of microscopic shells.

I have tried to give you very briefly a sketch of the essential points of Mr. Darwin's theory of evolution. But you should all read that marvel of patience, industry, clear insight, close reasoning, and grand honesty, the "Origin of Species." I have no time to give the arguments in its favor or to attempt to meet the objections which may arise in your minds. I ask you to believe only this much ; that the theory is accepted with practical unanimity by scientific men because it, and it alone, furnishes an explanation for the facts which they discover in their daily work. And this is the strongest proof of the truth of any accepted theory.

Inasmuch as it is accepted by all scientists and largely by the public, it is certainly worth your while to know whether it has any bearing on the great moral and religious questions which you are considering. And in these lectures I shall take for granted, what some scientists still doubt, that man also is a product of evolution. For the weight of evidence in favor of this view is constantly increasing, and seems already to strongly preponderate. Also I wish in these lectures to grant all that the most ardent evolutionist can possibly claim. Not that I would lower man's position, but I have a continually increasing respect for the so-called "lower animals."

Now if the theory of evolution be true, and really

only on this condition, life has had a history; and human history began ages before man's actual appearance on the globe, just as American history began to be fashioned by Anglo-Saxons, Danes, and Normans before they set foot even in England. We study history mainly to deduce its laws; and that knowing them we may from the past forecast the future, prepare for its emergencies, and avoid or wisely meet, its dangers. And we rely on these laws of history because they are the embodiment of ages of human experience.

Whatever be our system of philosophy we all practically rely on past experience and observation. Fire burns and water drowns. This we know, and this knowledge governs our daily lives, whatever be our theories, or even our ignorance, of the laws of heat and respiration. Now human history is the embodiment of the experience of the race; and we study it in the full confidence that, if we can deduce its laws, we can rely on racial experience certainly as safely as on that of the individual. Furthermore, if we can discover certain great movements or currents of human action or progress moving steadily on through past centuries, we have full confidence that these movements will continue in the future. The study of history should make us seers.

But the line of human progress is like a mountain road, veering and twisting, and often appearing to turn back upon itself, and having many by-roads, which lead us astray. If we know but a few miles of it we cannot tell whether it leads north or south or due west. But if from any mountain-top we can gain a clear bird's-eye view of its whole course, we easily distinguish the main road, its turns become quite insignificant, we see that it leads as directly as any engineering skill could locate it through the mountains to the fertile plains and rich harvests beyond.

Now our knowledge of the history of man covers so brief a period that we can scarcely more than hazard a guess as to the trend of human progress. Many of the most promising social movements are like by-roads which, at first less steep and difficult, end sooner or later against impassable obstacles. And even if there be a main line of march, advance seems to alternate with retreat, progress with retrogression. To illustrate further, the great waves rush onward only to fall back again, and we can hardly tell whether the tide is flowing or ebbing.

Yet already certain tendencies appear fairly clear. Governments tend to become democratic, if we define democracy as "any form of government in which the will of the people finds sovereign expression." The tendency of society seems to be toward furnishing all its members equality of opportunity to make the most of their natural endowments. But if we are convinced that these statements express even vaguely the tendency of human development in all its past history, we are confident that these tendencies will continue in the future for a period somewhat proportional to their time of growth in the past. If we are wise, we try to make our own lives and actions, and those of our fellows, conform to and advance them. Otherwise our lives will be thrown away.

But if the theory of evolution be true, human history is only the last page of the one history of all life. If we are to gain any adequate, true, extensive view of human progress, we must read more than this. We must take into account the history of man when he was not yet man. And if we believe in the future continuance of tendencies of a few centuries' growth, we shall rest assured of the permanence of tendencies which have grown and strengthened through the ages.

Our confidence in the results of historical study is therefore proportioned to the extent and thoroughness of the experience which they record, and to the time during which these laws can be proven to have held good. If I can make it even fairly probable that these laws, on obedience to which human progress and success seem to depend, are merely quoted from a grander code applicable to all life in all times, your confidence in them will be even greater. I trust I can prove to you that the animal kingdom has not drifted aimlessly at the mercy of every wind and tide and current of circumstance. I hope to show that along one line it has from the beginning through the ages held a steady course straight onward, and that deviation from this course has always led to failure or degeneration. From so vast a history we may hope to deduce some of the great laws of true success in life. Furthermore, if along this central line, at the head of which man stands, there always has been progress, we cannot doubt that future progress will be as certain, and perhaps far more rapid. In all the struggle of life we shall have the sure hope of success and victory ; if not for ourselves still for those who shall come after us. " We are saved by hope." And we may be confident that this hope will never make us ashamed.

Finally, even from our present knowledge of the

2

past progress of life we shall hope to catch hints at least that man's only path to his destined goal is the straight and narrow road pointed out in the Bible. If in this we are even fairly successful we shall find a relation and bond between the Bible and Science worthy of all consideration. And this is the only agreement which can ever satisfy us.

If I wished to bring before you a view of the development of man, I should best choose individuals or families from various periods of human history from the earliest times down to the present. I should try to tell you how they looked and lived. But if anyone should attempt to condense into three lectures such a history of even one line of the human race, you would probably think him insane. Even if he succeeded in giving a fairly clear view of the different stages, the successive stages would be so remote from one another, such vast changes would necessarily remain unnoticed or unexplained that you would hardly believe that they could have any genetic relation or belong to one developmental series.

But the history which I must attempt to condense for you is measured by ages, and the successive terms of the series will be indefinitely more remote from each other than the life and thoughts of Lincoln or Washington from those of our most primitive Aryan ancestor or of the rudest savage of the Stone Age. The series must appear exceedingly disconnected. Systems of organs will apparently spring suddenly into existence, and we shall have no time to trace their origin or earlier development. Even if we had an abundance of time many gaps would still remain; for the forms, which according to our theory must have occupied their place, have long since disappeared and left no trace nor sign. We have generally no conception at all of the amount of extermination and degeneration which have taken place in past ages.

I grant frankly that I do not believe that the forms which I have selected represent exactly the ancestors of man. They have all been more or less modified. I claim only that in the balance and relative development of their organic systems—muscular, digestive, nervous, etc.—they give us a very fair idea of what our ancestor at each stage must have been. But it is on this balance and relative development of the different systems, that is, whether an animal is more reproductive, digestive, or nervous, that my argument will in the main be based.

But if the older ancestors have so generally disappeared, and their surviving relatives have been so greatly modified, how can we make even a shrewd guess at the ancestry of higher forms? The genealogy of the animal kingdom has been really the study of centuries, although the earlier zoölogists did not know that this was to be the result of their labors. The first work of the naturalist was necessarily to classify the plants and animals which he found, and catalogue and tabulate them so that they might be easily recognized, and that later discovered forms might readily find a place in the system. Hypotheses and theories were looked upon with suspicion. "Even Linnæus," says Romanes, "was express in his limitations of true scientific work in natural history to the collecting and arranging of species of plants and animals." The question, "What is it?" came first; then, "How did it come to be what it is?" We are just awakening to the question, " Why this progressive system of forms, and what does it all mean ? "

Let us experiment a little in forming our own classification of a few vertebrates. We see a bat flying through the air. We mistake it for a bird. But a glance at it shows that it is a mammal. It is covered with hair. It has fore and hind legs. Its wings are membranes stretched between the fingers and along the sides of the body. It has teeth. It suckles its young. In all these respects it differs from birds. It differs from mammals only in its wings. But we remember that flying squirrels have a membrane stretching along the sides of the body and serving as a parachute, though not as wings. We naturally consider the wings as a sort of after-thought superinduced on the mammalian structure. We do not hesitate to call it a mammal.

The whale makes us more trouble ; it certainly looks remarkably like a fish. But the fin of its tail is horizontal, not vertical. Its front flippers differ altogether from the corresponding fins of fish ; their bones are the same as those occurring in the forelegs of mammals, only shorter and more crowded together. Later we find that it has lungs, and a heart with four chambers instead of only two, as in fish. The vertebræ of its backbone are not biconcave, but flat in front and behind. And, finally, we discover that it suckles its young. It, too, is in all its deep-seated characteristics a mammal. It is fish-like only in characteristics which it might easily have acquired in adaptation to its aquatic life. And there are other aquatic mammals, like the seals, in which these characteristics are much less marked. Their adaptation has evidently not gone so far.

Now the first attempts resulted in artificial classifications, much like our grouping of bats with birds and whales with fish. All animals, like coral animals and starfishes, whose similar parts were arranged in lines radiating from a centre, were united as radiates, however much they might differ in internal structure and grade of organization. But this radiate structure proved again to be largely a matter of adaptation.

Practically all animals having a heavy calcareous shell were grouped with the snails and oysters as mollusks. But the barnacle did not fit well with other mollusks. Its shell was entirely different. It had several pairs of legs ; and no mollusk has legs. The barnacle is evidently a sessile crab or better crustacean. Its molluscan characteristics were only skin-deep, evidently an adaptation to a mode of life like that of mollusks. The old artificial systems were based too much on merely external characteristics, the results of adaptation. When the internal anatomy had been thoroughly studied their groups had to be rearranged.

Reptiles and amphibia were at first united in one class because of their resemblance in external form. Our common salamanders look so much like lizards that they generally pass by this name. But the young salamander, like all amphibia, breathes by gills, its skeleton differs greatly from, and is far weaker than, that of the lizard, and there are important differences in the circulatory and other systems. Moreover, practically all amphibia differ from all reptiles in these respects. Evidently the fact that the alligator and many snakes and turtles (of which neither the young nor the embryos ever breathe by gills) live almost entirely in the water, is no better reason for classifying these with amphibia than to call a whale a fish, and not a mammal, because of its form and aquatic life.

When the comparative anatomy of fish, amphibia, and reptiles had been carefully studied it was evident that the amphibia stood far nearer the fish in general structure, while the higher reptiles closely approached birds. Then it was noticed that our common fish formed a fairly well-defined group, but that the ganoids, including the sturgeons, gar-pikes, and some others, had at least traces of amphibian characteristics. Such generalized forms, with the characteristics of the class less sharply marked, were usually by common consent placed at the bottom of the class. And this suited well their general structure, while in particular characteristics they were often more highly organized than higher groups of the same class.

The palæontologist found that the oldest fossil forms belonged to these generalized groups, and that more highly specialized forms—that is, those in which the special class distinctions were more sharply and universally marked—were of later geological origin. Thus the oldest fish were most like our present ganoids and sharks, though differing much from both. Our common teleost fish, like perch and cod, appeared much later. The oldest bird, the archæopteryx, had a long tail like that of a lizard, and teeth ; and thus stood in many respects almost midway between birds and reptiles. And most of the earliest forms were " comprehensive," uniting the characteristics of two or more later groups. Thus as the classification became more natural, based on a careful comparison of the whole anatomy of the animals, its order was found to coincide in general with that of geological succession.

Then the zoölogist began to ask and investigate how the animal grew in the egg and attained its definite form. And this study of embryology brought to light many new and interesting facts. Agassiz especially emphasized and maintained the universality of the fact that there was a remarkable parallelism between embryos of later forms and adults of old or fossil groups. The embryos of higher forms, he said, pass through and beyond certain stages of structure, which are permanent in lower and older members of the same group.

You remember that the fin on the tail of a fish is as a rule bilobed. Now the backbone of a perch or cod ends at a point in the end of the tail opposite the angle between the two lobes, without extending out into either of them. In the shark it extends almost to the end of the upper lobe. Now we have seen that sharks and ganoids are older than cod. In the embryo of the cod or perch the backbone has, at an early stage, the same position as in the shark or ganoid; only at a later stage does it attain its definite position.

So Agassiz says the young lepidosteus (a ganoid fish), long after it is hatched, exhibits in the form of its tail characters thus far known only among the fossil fishes of the Devonian period. The embryology of turtles throws light upon the fossil chelonians. It is already known that the embryonic changes of frogs and toads coincide with what is known of their succession in past ages. The characteristics of extinct genera of mammals exhibit everywhere indications that their living representatives in early life resemble them more than they do their own parents. A minute comparison of a young elephant with any mastodon will show this most fully, not only in the peculiarities of their teeth, but even in the proportion of their limbs, their toes, etc. It may therefore be considered as a general fact that the phases of development of all living animals correspond to the order of succession of their extinct representatives in past geological times. The above statements are quoted almost word for word from Professor Agassiz's "Essay on Classification." The larvæ of barnacles and other more degraded parasitic crustacea are almost exactly like those of crustacea in general. The embryos of birds have a long tail containing almost or quite as many vertebræ as that of archæopteryx. But most of these never reach their full development but are absorbed into the pelvis, or into the "ploughshare" bone supporting the tail feathers. Thus older forms may be said to have retained throughout life a condition only embryonic in their higher relatives. And the natural classification gave the order not only of geological succession but also of stages of embryonic development. Thus the system of classification improved continually, although more and more intermediate forms, like archæopteryx, were discovered, and certain aberrant groups could find no permanent resting-place.

But why should the generalized comprehensive forms stand at the bottom rather than the top of the systematic arrangement of their classes? Why should the system of classification coincide with the order of geologic occurrence, and this with the series of embryonic stages? Above all, why should the embryos of bird and perch form their tails by such a roundabout method? Why should the embryo of the bird have the tail of a lizard? No one could give any satisfactory explanation, although the facts were undoubted.

Mr. Darwin's theory was the one impulse needed to crystallize these disconnected facts into one comprehensible whole. The connecting link was everywhere common descent, difference was due to the continual variation and divergence of their ancestors. The classification, which all were seeking, was really the ancestral tree of the animal kingdom. Forms more generalized should be placed lower down on the ancestral tree, and must have had an earlier geological occurrence because they represented more nearly the ancestors of the higher. But this explains also the facts of embryonic development.

According to Mr. Darwin's theory all the species of higher animals have developed from unicellular ancestors. It had long been known that all higher forms start in life as single cells, egg and spermatozoon. And these, fused in the process of fertilization, form still a single cell. And when this single cell proceeds through successive embryonic stages to develop into an adult individual it naturally, through force of hereditary habit, so to speak, treads the same path which its ancestors followed from the unicellular condition to their present point of development. Thus higher forms should be expected to show traces of their early ancestry in their embryonic life. Older and lower adult forms should represent persistent embryonic stages of higher. It could not well be otherwise.

But the path which the embryo has to follow from the egg to the adult form is continually lengthening as life advances ever higher. From egg to sponge is, comparatively speaking, but a step ; it is a long march from the egg to the earthworm ; and the vertebrate embryo makes a vast journey. But embryonic life is and must remain short. Hence in higher forms the ancestral stages will often be slurred over and very incompletely represented. And the embryo may, and often does, shorten the path by "short-cuts" impossible to its original ancestor. Still it will in general hold true, and may be recognized as a law of vast importance, that any individual during his embryonic life repeats very briefly the different stages through which his ancestors have passed in their development since the beginning of life. Or, briefly stated, ontogenesis, or the embryonic development of the individual, is a brief recapitulation of phylogenesis, or the ancestral development of the phylum or group.

The illustration and proof of this law is the work of the embryologist. We have time to draw only one or two illustrations from the embryonic development of birds. We have already seen that the embryonic bird has the long tail of his reptilian ancestor. In early embryonic life it has gill-slits leading from the pharynx to the outside of the neck like those through which the water passes in the respiration of fish. The Eustachian tube and the canal of the external ear of man, separated only by the "drum," are nothing but such an old persistent gill-slit. No gills ever develop in these, but the great arteries run to them, and indeed to all parts of the embryo, on almost precisely the same general plan as in the adult fish. Only later is the definite avian circulation gradually acquired.

This law is even more strikingly illustrated in the embryonic development of the vertebral column and skull, if we had time to trace their development. And the development of the excretory system points to an ancestor far more primitive than even the fish. Our embryonic development is one of the very strongest evidences of our lowly origin.

Thus we have three sources of information for the study of animal genealogy. First, the comparative anatomy of all the different groups of animals; second, their comparative embryology; and third, their palæoutological history. Each source has its difficulties or defects. But taken all together they give us a genealogical tree which is in the main points correct, though here and there very defective and doubtful in detail. The points in which we are left most in doubt in regard to each ancestor are its modes of life and locomotion, and body form. But these may temporarily vary considerably without affecting to any great extent the general plan of structure and the line of development of the most important deep-seated organs.

I have chosen a line composed of forms taken from the comparative anatomical series. All such present existing forms have probably been modified during the lapse of ages. But I shall try to tell you when they have diverged noticeably from the structure of the primitive ancestor of the corresponding stage. It is much safer for us to study concrete, actual forms than imaginary ones, however real may have been the former existence of the latter. And, after all, their lateral divergence is of small account compared with the great upward and onward march of life, to the right and left of which they have remained stationary or retrograded somewhat, like the tribes which remained on the other side of Jordan and never entered the Promised Land.

To recapitulate : Our question is the Whence and the Whither of man. To this question the Bible gives a clear and definite answer. Can Science also give an answer, and is this in the main in accord with the answer of Scripture? Science can answer the question only by the historical method of tracing the history of life in the past and observing the goal toward which it tends. If the evolution theory be true, the record of human achievement and progress forms only one short chapter in the history of the ages. If from the records of man's little span of life on the globe we can deduce laws of history on whose truth we can rely, with how much greater confidence and certainty may we rely on laws which have governed all life since its earliest appearance?—always provided that such can be found.

Our first effort must therefore be to trace the great line of development through a few of its most characteristic stages from the simplest living beings up to man. This will be our work in the three succeeding lectures. And to these I must ask you to bring a large store of patience. Anatomical details are at best dry and uninteresting. But these dry facts of anatomy form the foundation on which all our arguments and hopes must rest.

But if you will think long and carefully even of anatomical facts, you will see in and behind them something more and grander than they. You will catch glimpses of the divinity of Nature. Most of us travel threescore years and ten stone-blind in a world of marvellous beauty. Why does the artist see so much more in every fence-corner and on every hill-side than we, set face to face with the grandest landscapes? Primarily, I believe, because he is sympathetic, and looks on Nature as a comrade as near and dear as any human sister and companion. As Professor Huxley has said, "they get on rarely together." She speaks to the artist; to us she is dumb, and ought to be, for we are boorishly careless of her and her teachings.

Nature, to be known, must be loved. And though you have all the knowledge of a von Humboldt, and do not love her, you will never understand her or her teachings. You will go through life with her, and yet parted from her as by an adamantine wall.

I do not suppose that the author of the book of Job had ever studied geology, or mineralogy, or biology, but read him, and see whether this old prince of scientific heroes had loved, and understood, and caught the spirit of Nature. And what a grand, free spirit it was, and what a giant it made of him. I do not believe that Paul ever had a special course of anatomy or botany. But if he had not pondered long and lovingly on the structure of his body, and the germination of the seed, he never could have written the twelfth and fifteenth chapters of the first letter to the Corinthians. And time fails to speak of David and all the writers of the Psalms, and of those heroic souls misnamed the "Minor" Prophets.

Study the teachings of our Lord. How he must have considered the lilies of the field, and that such a tiny seed as that of the mustard could have produced so great an herb, and noticed and thought on the thorns and the tares and the wheat, and watched the sparrows, and pondered and wondered how the birds were fed. All his teaching was drawn from Nature. And all the study in the world could never have taught him what he knew, if it had not been a loving and appreciative study.

There is one strange and interesting passage in John's Gospel, xv. 1: "I am the true vine." My father used to tell us that the Greek word ἀληθινή, rendered true, is usually employed of the genuine in distinction from the counterfeit, the reality in distinction from the shadow and image. Is not this perhaps the clew to our Lord's use of natural imagery? Nature was always the presentation to his senses of the divine thought and purpose. He studied the words of the ancient Scripture, he found the same words and teachings clearly and concretely embodied in the processes of Nature. The interpretation of the Parable of the Sower was no mere play of fancy to him; it was the genuine and fundamental truth, deeper and more real than the existence of the sower, the soil, and the seed. The spiritual truth was the substance; the tangible soil and seed really only the shadow. And thus all Nature was to him divine.

We all of us need to offer the prayer of the blind man, "Lord, that our eyes may be opened." Let us learn, too, from the old heathen giant, Antæus, who, after every defeat and fall, rose strengthened and vivified from contact with his mother Earth. You will experience in life many a desperate struggle, many a hard fall. There is at such times nothing in the world so strengthening, healing, and life-giving as the thoughts and encouragements which Nature pours into the hearts and minds of her loving disciples. She will set you on your feet again, infused with new life, filled with an unconquerable spirit, with unfaltering courage, and an iron will to fight once more and win. In every battle her inspiring words will ring in your ears. and she will never fail you. We may not see her deepest realities, her rarest treasures of thought and wisdom; but if we will listen lovingly for her voice, we may be assured that she will speak to us many a word of cheer and encouragement, of warning and exhortation. For, to paraphrase the language of the nineteenth Psalm, " She has no speech nor language, her voice is not heard. But her rule is gone out throughout all the earth, and her words to the end of the world."

CHAPTER II

THE first and lowest form in our ancestral series is the amoeba, a little fresh-water animal from $\frac{1}{500}$ to $\frac{1}{1000}$ of an inch in diameter. Under the microscope it looks like a little drop of mucilage. This semifluid, mucilaginous substance is the Protoplasm. Its outer portion is clear and transparent, its inner more granular. In the inner portion is a little spheroidal body, the nucleus. This is certainly of great importance in the life of the animal; but just what it does, or what is its relation to the surrounding protoplasm we do not yet know. There is also a little cavity around which the protoplasm has drawn back, and on which it will soon close in again, so that it pulsates like a heart. It is continually taking in water from the body, or the outside, and driving it out again, and thus aids in respiration and excretion. The animal has no organs in the proper sense of the word, and yet it has the rudiments of all the functions which we possess.

A little projection of the outer, clearer layer of protoplasm, a pseudopodium, appears ; into this the whole animal may flow and thus advance a step, or the projection may be withdrawn. And this power of change of form is a lower grade of the contractility of our muscular cells. Prick it with a needle and it contracts. It recognizes its food even at a microscopic distance; it appears therefore to feel and perceive. Perhaps we might say that it has a mind and will of its own. It is safer to say that it is irritable, that is, it reacts to stimuli too feeble to be regarded as the cause of its reaction. It engulfs microscopic plants, and digests them in the internal proto-plasm by the aid of an acid secretion. It breathes oxygen, and excretes carbonic acid and urea, through its whole body surface. Its mode of gaining the en-ergy which it manifests is therefore apparently like our own, by com-bustion of food mate-rial.

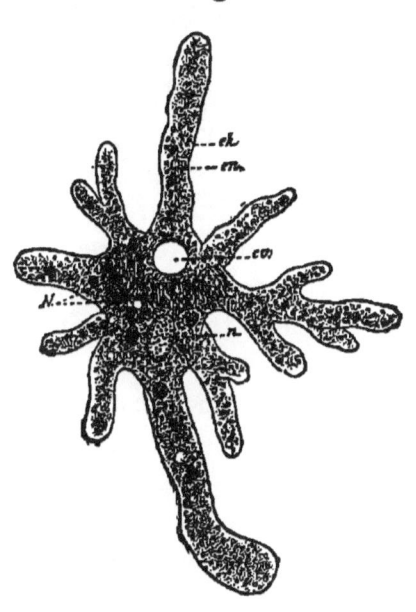

1. AMŒBA PROTEUS. HERTWIG, FROM LEIDY.

ek, ectosarc ; *en,* endosarc ; *N,* food par-ticles ; *n,* nucleus ; *cv,* contractile vesicle.

It grows and reaches a certain size, then con-stricts itself in the mid-dle and divides into two. The old amœba has divided into two young ones, and there is no parent left to die, and death, except by violence, does not occur. But this absence of death in other rather distant relatives of the amœba, and probably in the amœba itself, holds true only provided that, after a series of self-divisions, reproduction takes place after another mode. Two rather small and weak individuals fuse together in one animal of renewed vigor, which soon divides into two

3

larger and stronger descendants. We have here evidently a process corresponding to the fertilization of the egg in higher animals; yet there is no egg, spermatozoon, or sex.

It is a little mass of protoplasm containing a nucleus, and corresponds, therefore, to one of the cells, most closely to the egg-cell or spermatozoon of higher animals. If every living being is descended from a single cell, the fertilized egg, it is not hard to believe that all higher animals are descended from an ancestor having the general structure or lack of structure of the amœba.

But is the amœba really structureless? Probably it has an exceedingly complex structure, but our microscopes and technique are still too imperfect to show more than traces of it. Says Hertwig: "Protoplasm is not a single chemical substance, however complicated, but a mixture of many substances, which we must picture to ourselves as finest particles united in a wonderfully complicated structure." Truly protoplasm is, to borrow Mephistopheles' expression concerning blood, a "quite peculiar juice." And the complexity of the nucleus is far more evident than that of the protoplasm. Is protoplasm itself the result of a long development? If so, out of what and how did it develop? We cannot even guess. But the beginning of life may, apparently must, have been indefinitely farther back than the simplest now existing form. The study of the amœba cannot fail to raise a host of questions in the mind of any thoughtful man.

As we have here the animal reduced, so to speak, to lowest terms, it may be well to examine a little more closely into its physiology and compare it briefly with our own.

The amœba eats food as we do, but the food is di-· gested directly in the internal protoplasm instead of in a stomach; and once digested it diffuses to all parts of the cell; here it is built up into compounds of a more complex structure, and forms an integral part of the animal body. The dead food particle has been transformed into living protoplasm, the continually repeated miracle of life. But it does not remain long in this condition. In contact with the oxygen from the air it is soon oxidized, burned up to furnish the energy necessary for the motion and irritability of the body. We are all of us low-temperature engines. The digestive function exists in all animals merely to bring the food into a soluble, diffusible form, so that it can pass to all parts of the body and be used for fuel or growth. In our body a circulatory system is necessary to carry food and oxygen to the cells and to remove their waste. For most of our cells lie at a distance from the stomach, lungs, and kidney. But in a small animal the circulatory system is often unnecessary and fails. Breathing and excretion take place through the whole surface of the body. The body of the frog is devoid of scales, so that the blood is separated from the surrounding water only by a thin membrane, and it breathes and excretes to a certain extent in the same way.

But another factor has to be considered. If we double each dimension of our amœba, we shall increase its surface four times, its mass eight-fold. Now the power of absorbing oxygen and excreting waste is evidently proportional to the excretory and respiratory surface, and much the same is true of digestion. But the amount of oxygen required, and of waste to be removed is proportional to the mass; for every par-

ticle of protoplasm requires food and oxygen, and produces waste. The particles of protoplasm in our new,
larger amœba can therefore receive only half as much
oxygen as before, and rid themselves of their waste
only half as fast. There is danger of what in our
bodies would be called suffocation and blood-poisoning.
The amœba having attained a certain size meets this
emergency by dividing into two small individuals, the
division is a physical adaptation. But the many-celled
animal cannot do this ; it must keep its cells together.
It gains the additional surface by folding and plaiting.
And the complicated internal structure of higher animals is in its last analysis such a folding and plaiting
in order to maintain the proper ratio between the exposed surface of the cells and their mass. And each
cell in our bodies lives in one sense its own individual life, only bathed in the lymph and receiving from
it its food and oxygen instead of taking it from the
water.

But in another sense the cells of our body live an
entirely different life, for they form a community.
Division of labor has taken place between them, they
are interdependent, correlated with one another, subject
therefore to the laws of the whole community or organism. There are many respects in which it is impossible to compare Robinson Crusoe with a workman in
a huge watch factory ; yet they are both men.

Both the amœba and we live in the closest relation
to our environment, and conformity to it is evidently
necessary : life has been defined as the adjustment of
internal relations to external conditions. We continually take food, use it for energy and growth, and return
the simpler waste compounds. We are all of us, as

Professor Huxley has said, "whirlpools on the surface of Nature;" when the whirl of exchange of particles ceases we die. We have seen that the fusion of two amœbæ results in a new rejuvenated individual. Why is a mixture of two protoplasms better than one ? We can frame hypotheses; we know nothing about it. What of the mind of the amœba ? A host of questions throng upon us and we can answer no one of them. All the great questions concerning life confront us here in the lowest term of the animal series, and appear as insoluble as in the highest.

Our second ancestral form is also a fresh-water animal, the hydra. This is a little, vase-shaped animal, which usually lives attached to grass-stems or sticks, but has the power to free itself and hang on the surface of the water or to slowly creep on the bottom. The mouth is at the top of the vase, and the simple, undivided cavity within the vase is the digestive cavity. Around the mouth is a ring of from four to ten hollow tentacles, whose cavities communicate freely underneath with the digestive cavity. Not only is food taken in at the mouth, but indigestible material is thrown out here. The animal may thus be compared to a nearly cylindrical sack with a circle of tubes attached to it above. The body consists of two layers of cells, the ectoderm on the outside and the entoderm lining the digestive cavity. Between these two is a structureless, elastic membrane, which tends to keep the body moderately expanded.

The food is captured by the tentacles; but digestion takes place only partially in the digestive cavity, for each surrounding cell engulfs small particles of food and digests them within itself. The entodermal cells behave in this respect much like a colony of amœbæ. The cells of both layers have at their bases long muscular fibrils, those of the ectodermal cells running longitudinally, those of the entoderm transversely. The animal can thus contract its body in both directions, or, if the body contain water and the transverse muscles are contracted, the pressure of the water lengthens the body and tends to extend the tentacles.

On the outside of the elastic membrane, just beneath the ectoderm, is a plexus or cobweb of nervous cells and fibrils. As in every nervous system, three elements are here to be found. 1. An afferent or sensory nerve-fibril, which under adequate stimulus is set in vibration by some cell of the epidermis or ectoderm, which is therefore called a sensory cell. 2. A central or ganglion cell, which receives the sensory impulse, translates it into consciousness, and is the seat of whatever powers of perception, thought, or will the animal possesses. This also gives rise to the efferent or motor impulses, which are conveyed by (3) a motor fibril to the corresponding muscle, exciting its contraction. But there are also nerve-fibrils connecting the different ganglion cells, so that they may act in unison. In the higher animals we shall find these central or ganglion cells condensed in one or a few masses or ganglia. But here they are scattered over the whole surface of the elastic supporting membrane.

The reproductive organs for the production of eggs and spermatozoa form little protuberances on the outside of the body below the tentacles. But hydra reproduces mostly by budding; new individuals growing out of the side of the old one, like branches from the trunk of a tree, but afterward breaking free and leading an independent life. There are special forms of cells besides those described; nettle cells for capturing food, interstitial cells, etc., but these do not concern us.

The distance from the single-celled amœba to hydra is vast, probably really greater than that between any other successive terms of our series. It may therefore be useful to consider one or two intermediate forms and the parallel embryonic stages of higher animals, and to see how the higher many-celled animal originates from the unicellular stage.

The amœba is an illustration of a great kingdom of similar, practically unicellular forms, which have played no unimportant part in the geological history of the globe. These are the protozoa. They include, first of all, the foraminifera, which usually have shells composed of carbonate of lime. These shells, settling to the bottom of the ocean, have accumulated in vast beds, and when compacted and raised above the surface, form chalk, limestone, or marble, according to the degree and mode of their hardening.

 The protozoa include also the flagellata, a great, very poorly defined mass of forms occupying the boundary between the plant and animal kingdoms. They are usually unicellular, and their protoplasm is surrounded by a thin, structureless membrane. This prevents their putting out pseudopodia as organs of motion. Instead of these they have at one end of the ovoid or pear-shaped body a long, whiplash-like process or thread, a flagellum, and by swinging this they propel themselves through the water. These flagellata seem to have a rather marked tendency to form colonies. The first individual gives rise to others by

division. But the division is not complete ; the new
individuals remain connected by the undivided rear
end of the body. And such a colony may come to
contain a large number of individuals.

Such a colony is represented by magosphæra. This
is a microscopic globular form, discovered by Professor
Haeckel on the coast of Norway. It consists of a

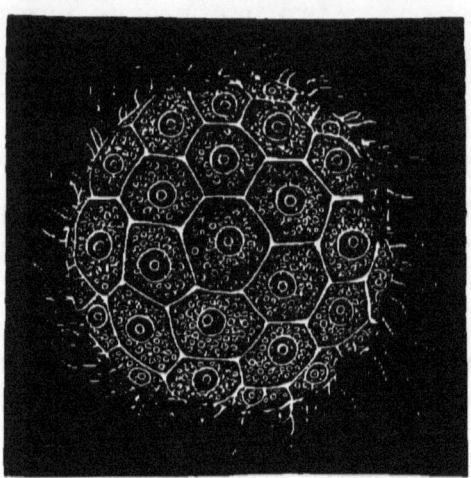

large number of
conical or pear-
shaped individual
cells, whose apices
are turned toward
the centre of the
sphere. The cells
are cemented to-
gether by a mucila-
ginous substance.
Around their ex-
posed larger ends,
which form the sur-
face of the sphere,
are rows of flagel-

2. MAGOSPHÆRA PLANULA. LANG, FROM HAECKEL.

la, by whose united action the colony rolls through the
water. After a time each individual absorbs its flagella, the colony is broken up, the different individuals
settle to the bottom, and each gives rise by division to
a new colony. This group of cells may be considered
as a colony or as an individual. Each term is de-
fensible.

Volvox is also a spheroidal organism, composed often of a very large number of flagellated cells. But it
differs from magosphæra in certain important respects.
In the first place its cells have chlorophyl, the green

coloring matter of plants. It lives therefore on un-organized fluid nourishment, carbon dioxide, nitrates, etc. It is a plant. But certain characteristics render it probable that it once lived on solid food and was therefore an animal. For where almost the sole difference between plants and animals is in the fluid or solid character of their food, a change from the one form into the other is not as difficult or improbable as one might naturally think. And plants and animals are here so near together, and travelling by roads so nearly parallel, that, even if volvox never was an animal, it might still serve very well to illustrate a stage through which animals must have passed.

The cells of volvox do not form a solid mass, but have arranged themselves in a single layer on the outer surface of the sphere. For a time, under favorable circumstances, volvox reproduces very much like magosphæra, and each cell can give rise to a new, many-celled individual. But after a time, especially under unfavorable circumstances, a new mode of reproduction appears. Certain cells withdraw from the outer layer into the interior of the colony. Here they are nourished by the other cells and develop into true reproductive elements, eggs and spermatozoa. Fertilization, that is, the union of egg and spermatozoon, or mainly of their nuclei, takes place; and the fertilized egg develops into a new organism. But the other cells, which have been all the time nourishing these, seem now to lack nutriment, strength, or vitality to give rise to a new colony. They die.

We find thus in volvox division of labor and corresponding difference of structure or differentiation; certain cells retain the power of fusing with other corresponding cells, and thus of rejuvenescence and of giving rise to a new organism. And these cells, forming a series through all generations, are evidently immortal like the protozoa. Natural death cannot touch them. These are the reproductive cells. The other cells nourish and transport them and carry on the work of excretion and respiration. These latter correspond practically to our whole body. We call them somatic cells. In volvox they are entirely subservient to, and exist for, the reproductive cells, and die when they have completed their service of these. The body is here only a vehicle for ova. Furthermore, in volvox there has arisen such an interdependence of cells that we can no longer speak of it as a colony. The colony has become an individual by division of labor and the resulting differentiation in structure.

But hydra gives us but a poor idea of the cœlenterata, to which kingdom it belongs. The higher cœlenterata have nearly or quite all the tissues of higher animals—muscular, connective, glandular, etc. And by tissues we mean groups of cells modified in form and structure for the performance of a special work or function. The protozoa developed the cell for all time to come, the cœlenterata developed the tissues which still compose our bodies. But they had them mainly in a diffuse form. A sort of digestive and reproductive system they did possess. But the work of arranging these tissues and condensing them into compact organs was to be done by the next higher group, the worms.

Let us now take a glance at certain stages of embryonic development which correspond to these earliest ancestral forms. We should expect some such correspondence from the fact already stated that the embryonic development of the individual is a brief recapitulation of the ancestral development of the species or larger group. The egg of the lowest vertebrate, amphioxus, shows these changes in a simple and apparently primitive form.

The fertilized egg of any animal consists of a single cell, a little mass of protoplasm containing a nucleus and surrounded by a structureless membrane. The egg is globular. The nucleus undergoes certain very peculiar, still but little understood, changes and divides into two. The protoplasm also soon divides into two masses clustering each around its own nucleus. The plane of division will be marked around the outside by a circular furrow, but the cells will still remain united by a large part of the membrane which bounds their adjacent, newly formed, internal faces.

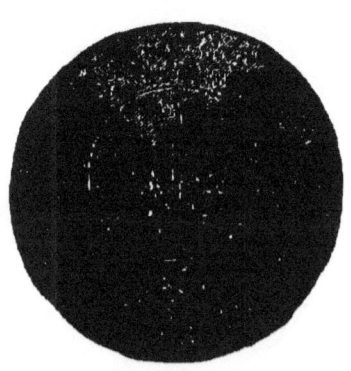

3. IMMATURE EGG-SHELL FROM OVARY OF ECHINODERM. HATSCHEK, FROM HERTWIG.

Let us suppose that the egg lay so that the first plane of division was vertical and extending north and south. Each cell or half of the egg will divide into two precisely as before. The new plane of division will be vertical, but extending east and west. Each plane passes through the centre of the egg, and the four cells are of the same form and size, like much-rounded quarters of an orange. The third plane will lie horizontal or equatorial, and will divide each of these quarters into an upper and lower octant. The cells keep on dividing rapidly, the eight form sixteen, then thirty-two, etc. The sharp angle by which the cells met at the centre has become rounded off, and has left a little space, the segmentation cavity, filled with fluid in the middle of the embryo. The cells continue to press or be crowded away from the centre and form a layer one cell deep on the surface of the sphere.

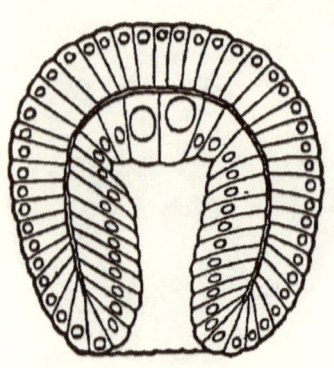

4. GASTRULA. HATSCHEK, FROM HERTWIG.

Outer layer is the ectoderm ; inner layer, the entoderm ; internal cavity, the archenteron ; mouth of cavity, blastopore.

This embryo, resembling a hollow rubber ball filled with fluid, is called a blastosphere. It corresponds in structure with the fully developed volvox, except, of course, in lacking reproductive cells.

If the rubber ball has a hole in it so that I can squeeze out the water, I can thrust the one-half into the other, and change the ball into a double-walled cup. A similar change takes place in the embryo. The cells of the lower half of the blastosphere are slightly larger than those of the upper half. This lower hemisphere flattens and then thrusts itself, or is invaginated, into the upper hemisphere of smaller cells and forms its lining. This cup-shaped embryo is called the gastrula. The cup deepens somewhat and becomes ovoid. Take a boiled egg, make a hole in the smaller end and remove the yolk, and you have a passable model of a gastrula. The shell corresponds to the ectoderm or outer layer of smaller cells; the layer of " white " represents the entoderm or lining of larger cells. The space occupied by the yolk corresponds to the archenteron or primitive digestive cavity ; and the opening at the end to the primitive mouth or blastopore. Ectoderm and entoderm unite around the mouth. Both the blastosphere and gastrula often swim freely by flagella.

You can hardly have failed to notice how closely the gastrula corresponds to a hydra, and many facts lead us to believe that the still earlier ancestor of the hydra was free swimming, and that the tentacles are a later development correlated with its adult sessile life. Yet we must not forget that the hydra is even now not quite sessile, it moves somewhat. And our ancestor was almost certainly a free swimming gastraea, or hypothetical form corresponding in form and structure to the gastrula. The ancestor of man never settled down lazily into a sessile life.

But how is an adult worm or vertebrate formed out of such a gastrula ? To answer this would require a course of lectures on embryology. But certain changes interest us. Between the ectoderm and entoderm of the gastrula, in the space occupied by the supporting membrane of hydra, a new layer of cells, the mesoderm, appears. This has been produced by the rapid growth and reproduction of certain cells of the entoderm which have migrated, so to speak, into this new position. In higher forms it becomes of continually greater importance, until finally nearly all the organs of the body develop from it. In our bodies only the lining of the mid-intestine and of its glands has arisen from the entoderm. And only the epidermis, or outer layer of our skin, and the nervous system and parts of our sense-organs have arisen from the ectoderm. But our mid-intestine is still the greatly elongated archenteron of the gastrula.

We may therefore compare the hydra or gastrula to a little portion of the lining of the human mid-intestine covered with a little flake of epidermis. This much the hydra has attained. But our bones and muscles and blood-vessels all come from the mesoderm by folding, plaiting, and channelling, and division of labor resulting in differentiation of structure. Of all true mesodermal structures the hydra has actually none, but in the ectodermal and entodermal cells he has the potentiality of them all. We must now try to discover how these potentialities became actualities in higher forms.

The third stage in our ancestral series is the turbellarian. This is a little, flat, oval worm, varying greatly in size in different species, and found both in fresh and salt water. Some would deny that this worm belonged in our series at all. But, while doubtless considerably modified, it has still retained many characteristics almost certainly possessed by our primitive bilateral ancestor. The different parts of hydra were arranged like those of most flowers, around one main vertical axis; it was thus radiate in structure, having neither front nor rear, right nor left side. But our little turbellaria, while still without a head, has one end which goes first and can be called the front end. The upper or dorsal surface is usually more colored with pigment cells than the lower or ventral surface, on which is the mouth. It has also a right and left side. It is thus bilateral.

The gastræa swam by cilia, little eyelash - like

processes which urge the animal forward like a myriad of microscopic oars. In our bodies they are sometimes used to keep up a current, *e.g.*, to remove foreign particles from the lungs. The turbellaria is still covered with cilia, probably an inheritance from the gastræa; for, while in smaller forms they may still be the principal means of locomotion, in larger ones the muscles are beginning to assume this function and the animal moves by writhing. The bilateral symmetry has arisen in connection with this mode of locomotion and is thus a mark of important progress.

In the turbellaria we find for the first time a true body-wall distinct from underlying organs. The outer layer of this is a ciliated epithelium or layer of cells. Under this an elastic membrane may occur. Then come true body muscles, running transversely, longitudinally and dorso-ventrally. Between the external transverse and the internal longitudinal layers we often find two muscular layers whose fibres run diagonally. The body is well provided with muscles, but their arrangement is still far from economical or effective.

Within the body-wall is the parenchym. This is a spongy mass of connectile tissue in which the other organs are embedded. The mouth lies in the middle, or near the front of the ventral surface. The intestine varies in form, but is provided with its own layers of longitudinal and transverse muscles, and usually has paired pouches extending out from it into the body parenchym. These seem to distribute the dissolved nutriment; hence the whole cavity is still often called a gastro-vascular cavity as serving both digestion and circulation. There is no anal opening, but indigestible material is still cast out through the mouth.

The animal can gain sufficient oxygen to supply its muscles and nerves, which are the principal seats of combustion, through the external surface. It has, therefore, no special respiratory organs. But the waste matter of the muscles cannot escape so easily, for these are becoming deeper seated. Hence we find an excretory system consisting of two tubes with many branches in the parenchym, and discharging at the rear end of the body. This again is a sign that the muscles are becoming more important, for the excretory system is needed mainly to remove their waste. These tubes may be only greatly enlarged glands of the skin.

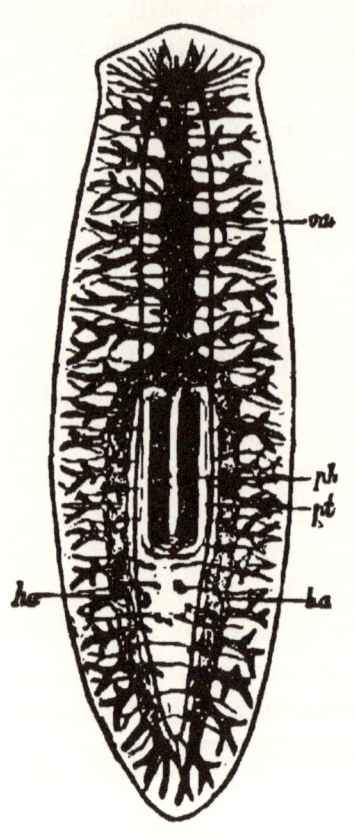

5. TURBELLARIAN. LANG.

va and *ha*, front and rear branches of gastro-vascular cavity; *ph,* pharynx. The dark oval with fine branches represents the nervous system.

The nervous system consists of a plexus of fibres and cells, the cells originating impulses and the fibres conveying them. But this much was present in hydra also. Here the front end of the body goes foremost and is continually coming in contact with new conditions. Here the lookout for food and danger must be kept. Hence, as a result of constant exercise, or selection, or both, the nerve-plexus has thickened at this point into a little compact mass of cells and fibres called a ganglion. And because this ganglion throughout higher forms usually lies over the œsophagus, it is called the supra-œsophageal ganglion. This is the first faint and dim prophecy of a brain, and it sends its nerves to the front end of the body. But there run from it to the rear end of the body four to eight nerve-cords, consisting of bundles of nerve-threads like our nerves, but overlaid with a coating of ganglion cells capable of originating impulses. These cords are, therefore, like the plexus from which they have condensed, both nerves and centres ; differentiation has not gone so far as at the front of the body. Sense organs are still very ʀrudimentary. Special cells of the skin have been modified into neuro - epithelial cells, having sensory hairs protruding from them and nerve - fibrils running from their bases.

In a very few turbellaria we find otolith vesicles.

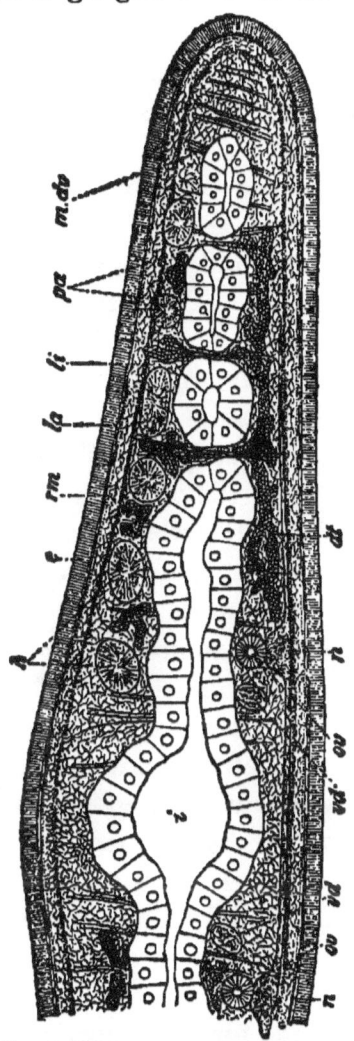

6. CROSS-SECTION OF TURBELLARIAN. HATSCHEK, FROM JIJIMA.

e, external skin ; *rm*, lateral muscles ; *la* and *li*, longitudinal muscles ; *mdv*, dorso-ventral muscles ; *pa*, parenchyma ; *h*, testʼcle ; *ov*, oviduct ; *dt*, yolk-gland ; *n*, ventral nerve ; *i*, gastro-vascular cavity.

These are little sacks in the skin, lined with neuro-epithelial cells and having in the middle a little con-cretion of carbonate of lime hung on rather a stiffer hair, like a clapper in a bell. Such organs serve in higher animals as organs of hearing, for the sensory hairs are set in vibration by the sound-waves. It is quite as probable that they here serve as organs for feeling the slightest vibrations in the surrounding water, and thus giving warning of approaching food or danger. The animal has also eyes, and these may be very numerous. They are not able to form images of external objects, but only of perceiving light and the direction of its source. A little group of these eyes lies directly over the brain, near the front end of the body; the others are distributed around the front or nearly the whole margin of the body.

The turbellaria, doubtless, have the sense of smell, although we can discover no special olfactory organ. This sense would seem to be as old as protoplasm itself.

This distribution of the eyes around a large portion of the margin, and certain other characteristics of the adult structure and of the embryonic development, are very interesting, as giving hints of the development of the turbellaria from some radiate ancestor. The mouth is in a most unfavorable position, in or near the middle of the body, rarely at the front end, as the animal has to swim over its food before it can grasp it. The animal only slowly rids itself of old disadvanta-geous form and structure and adapts itself completely to a higher mode of life.

By far the most highly developed system in the body is the reproductive. It is doubtful whether any animal, except, perhaps, the mollusk, has as complicated and highly developed reproductive organs. By markedly higher forms they certainly grow simpler.

And here we must notice certain general considerations. We found that reproduction in the amœba could be defined as growth beyond the limit normal to the individual. This form of growth benefits especially the species. The needs and expenses of the individual will therefore first be met and then the balance be devoted to reproduction. Now the income of the animal is proportional to its surface, its expense to its mass and activity. And the ratio of surface to mass is most favorable in the smallest animals.* Hence, smaller animals, as a rule, increase faster than larger ones ; and this is only one illustration of the fact that great size in an animal is anything but an unmixed advantage to its possessor. But muscles and nerves are the most expensive systems ; here most of the food is burned up. Hence energetic animals have a small balance remaining. Now the turbellarian is small and sluggish, with a fair digestive system. With a great amount of nutriment at its disposal the reproductive system came rapidly to a high development, and relatively to other organs stands higher than it almost ever will again.

It is only fair to state that good authorities hold that so primitive an animal could not originally have had so highly developed a system, and that this characteristic must be acquired, not ancestral.

That certain portions of it may be later developments may be not only possible but probable. But anyone who has carefully studied the different groups of worms, will, I think, readily grant that in the stage

* Cf. p. 35.

of these flat worms reproduction was the dominant function, which had most nearly attained its possible height of development. From this time on the muscular and nervous systems were to claim an ever-increasing share of the nutriment, and the balance for reproduction is to grow smaller.

At the close of this lecture I wish to describe very briefly a hypothetical form. It no longer exists ; perhaps it never did. But many facts of embryology and comparative anatomy point to such a form as a very possible ancestor of all forms higher than flat worms, viz., mollusks, arthropods, and vertebrates.

It was probably rather long and cylindrical, resembling a small and short earthworm in shape. The skin may have been much like that of turbellaria. Within this the muscles run in only two directions— longitudinally and transversely. Between these and the intestine is a cavity—the perivisceral cavity—like that of our own bodies, but filled with a nutritive fluid like our lymph. This cavity seems to have developed by the expansion and cutting off of the paired lateral outgrowths of the digestive system of some old flat worm. But other modes of development are quite possible. The intestine has now an anal opening at or near the rear end of the body. The food moves only from front to rear, and reaches each part always in a certain condition. Digestion proper and absorption have been distributed to different cells, and the work is better done. Three portions can be readily distinguished : fore-intestine with the mouth, mid-intestine, as the seat of digestion and absorption, and hind-intestine, or rectum, with the anal opening. The front and hind-intestine are lined with infolded outer skin.

The nervous system consists of a supra-œsophageal ganglion with four posterior nerve-cords—one dorsal, two lateral, and one (or perhaps two) ventral. There were probably also remains of the old plexus, but this is fast disappearing. The excretory system consists of a pair of tubes discharging through the sides of the body-wall, and having each a ciliated, funnel-shaped opening in the perivisceral cavity. These have received the name of nephridia. Through these also the eggs and spermatozoa are discharged. The reproductive organs are modified patches of the peritoneum, or lining of the perivisceral cavity.

The number of muscles or muscular layers has been reduced in this animal. But such a reduction in the number of like parts in any animal is a sign of progress. And the longitudinal muscles have increased in size and strength, and the animal moves by writhing. Such a worm has the general plan of the body of the higher forms fairly well, though rudely, sketched. Many improvements will come, and details be added. But the rudiments of the trunk of even our own bodies are already visible. Head, in any proper sense of the term, and skeleton are still lacking; they remain to be developed.

And yet, taking the most hopeful view possible concerning the animal kingdom, its prospects of attaining anything very lofty seem at this point poor. Its highest representative is a headless trunk, without skeleton or legs. It has no brain in any proper sense of the word, its sense-organs are feeble ; it moves by writhing. Its life is devoted to digestion and reproduction. Whatever higher organs it has are subsidiary to these lower functions. And yet it has taken ages on ages to develop this much. If *this* is the highest visible result of ages on ages of development, what hope is there for the future ? Can such a thing be the ancestor of a thinking, moral, religious person, like man ? "That is not first which is spiritual, but that which is natural (animal, sensuous) ; and afterward that which is spiritual." First, in order of time, must come the body, and then the mind and spirit shall be enthroned in it. The little knot of nervous material which forms the supra-œsophageal ganglion is so small that it might easily escape our notice ; but it is the promise of an infinite future. The atom of nervous power shall increase until it subdues and dominates the whole mass.

CHAPTER III

IN tracing the genealogy of any American family it is often difficult or impossible to say whether a certain branch is descended from John Oldworthy or his cousin or second cousin. In the latter cases to find the common ancestor we must go back to the grand-father or great-grandfather. The same difficulty, but greatly enhanced, meets us when we try to make a genealogical tree of the animal kingdom. Thus it seems altogether probable that all higher forms are descended from an ancestor of the same general struct-ure and grade of organization as the turbellaria, al-though probably free swimming, and hence with some-what different form and development, especially of the muscular system. It seems to me altogether probable that all, except possibly Mollusca, are descended from a common ancestor closely resembling the schematic worm last described. Some would, however, maintain that they diverged rather earlier than even the turbel-laria ; others after the schematic worm, if such ever existed. As far as our argument is concerned it makes little difference which of these views we adopt.

From our turbellaria, or possibly from some even more primitive ancestor, many lines diverged. And this was to be expected. The cœlenterata, as we saw in hydra, had developed rude digestive and reproductive systems. The higher groups of this kingdom had developed all, or nearly all, the tissues used in building the bodies of higher animals—muscular, reproductive, connectile, glandular, nervous, etc. But these are mostly very diffuse. The muscular fibrils of a jellyfish are mostly isolated or parallel in bands, rarely in compact well-defined bundles. The tissues have generally not yet been moulded into compact masses of definite form. There are as yet very few structures to which we can give the name of organs. To form organs and group them in a body of compact definite form was the work pre-eminently of worms. The material for the building was ready, but the architecture of the bilateral animal was not even sketched. And different worms were their own architects, untrammelled by convention or heredity, hence they built very different, sometimes almost fantastic, structures.

We must remember, too, the great age of this group. They are present in highly modified forms in the very oldest palæozoic strata, and probably therefore came into existence as the first traces of continental areas were beginning to rise above the primeval ocean. They are literally " older than the hills." They were exposed to a host of rapidly changing conditions, very different in different areas. This prepares us for the fact that the worms represent a stage in animal life corresponding fairly well to the Tower of Babel in biblical history. The animal kingdom seems almost to explode into a host of fragments. Our genealogical tree fairly bristles with branches, but the branches do not seem to form any regular whorls or spirals. Few of them have developed into more than feeble growths. They now contain generally but few species. Many of them are largely or entirely parasitic, and in connection with this mode of life have undergone modifications and degeneration which make it exceedingly difficult to decipher their descent or relationships.

Four of these branches have reached great prominence in numbers and importance. One or two others were formerly equally numerous and have since become almost extinct; so the brachiopoda, which have been almost entirely replaced by mollusks. The same may very possibly be true of others. For of the amount of extinction of larger groups we have generally but an exceedingly faint conception. Indeed in this respect the worms have been well compared to the relics which fill the shelves of one of our grandmother's china-closets.

The four great branches are the echinoderms, mollusks, articulates, and vertebrates. The echinoderms, including starfishes, sea-urchins, and others straggled early from the great army. We know as yet almost nothing of their history; when deciphered it will be as strange as any romance. The vertebrates are of course the most important line, as including the ancestors of man. But we must take a little glance at mollusks, including our clams, snails, and cuttle-fishes; and at the articulates, including annelids and culminating in insects. The molluscan and articulate lines, though divergent, are of great importance to us as throwing a certain amount of light on vertebrate development; and still more as showing how a certain line of development may seem, and at first really be, advantageous, and still lead to degeneration, or at best to but partial success.

When we compare the forms which represent fairly well the direction of development of these three lines, a snail or a clam with an insect and a fish, we find clearly, I think, that the fundamental anatomical difference lies in the skeleton; and that this resulted from, and almost irrevocably fixed, certain habits of life.

We may picture to ourselves the primitive ancestor of mollusks as a worm having the short and broad form of the turbellaria, but much thicker or deeper vertically. A fuller description can be found in the "Encyclopædia Britannica," Art., Mollusca. It was hemiovoid in form. It had apparently the perivisceral cavity and nephridia of the schematic worm, and a circulatory system. In this latter respect it stood higher than any form which we have yet studied. Its nervous system also was rather more advanced. It had apparently already taken to a creeping mode of life and the muscles of its ventral surface were strongly developed, while its exposed and far less muscular dorsal surface was protected by a cap-like shell covering the most important internal organs. But the integument of the whole dorsal surface was, as is not uncommon in invertebrates, hardening by the deposition of carbonate of lime in the integument. And this in time increased to such an extent as to replace the primitive, probably horny, shell.

Into the anatomy of this animal or of its descendants we have no time to enter, for here we must be very brief. We have already noticed that the most important viscera were lodged safely under the shell. And as these increased in size or were crowded upward by the muscles of the creeping disk, their portion of the body grew upward in the form of a "visceral hump." Apparently the animal could not increase much in length and retain the advantage of the protection of the shell; and the shell was the dominating structure. It had entered upon a defensive campaign. Motion, slow at the outset, became more difficult, and the protection of the shell therefore all the more necessary. The shell increased in size and weight and motion became almost impossible. The snail represents the average result of the experiment. It can crawl, but that is about all; it is neither swift nor energetic. Even the earthworm can outcrawl it. It has feelers and eyes, and is thus better provided with sense-organs than almost any worm. It has a supra-œsophageal ganglion of fair size.

The clams and oysters show even more clearly what we might call the logical results of molluscan structure. They increased the shell until it formed two heavy "valves" hanging down on each side of the body and completely enclosing it. They became almost sessile, living generally buried in the mud and gaining their food, consisting mostly of minute particles of organic matter, by means of currents created by cilia covering the large curtain-like gills. Their muscular system disappeared except in the ploughshare-shaped "foot" used mostly for burrowing, and in the muscles for closing the shell. That portion of the body which corresponds to the head of the snail practically aborted with nearly all the sense-organs. The nervous system degenerated and became reduced to a rudiment. They had given up locomotion, had withdrawn, so to speak, from the world; all the sense they needed was just enough to distinguish the particles of food as they swept past the mouth in the current of water. They have an abundance of food, and "wax fat."

The clam is so completely protected by his shell and the mud that he has little to fear from enemies. They have increased and multiplied and filled the mud. " Requiescat in pace."

But zoölogy has its tragedies as well as human history. Let us turn to the development of a third molluscan line terminating in the cuttle-fishes. The ancestors of these cephalopods, although still possessed of a shell and a high visceral hump, regained the swimming life. First, apparently, by means of fins, and then by a simple but very effective use of a current of water, they acquired an often rapid locomotion. The highest forms gave up the purely defensive campaign, developed a powerful beak, led a life like that of the old Norse pirates, and were for a time the rulers and terrors of the sea. With their more rapid locomotion the supra-œsophageal ganglion reached a higher degree of development, and it was served by sense-organs of great efficiency. They reduced the external shell, and succeeded, in the highest forms, of almost ridding themselves of this burden and encumbrance. Traces of it remain in the squids, but transformed into an internal quill-like, supporting, not defensive, skeleton. They have retraced the downward steps of their ancestors as far as they could. And the high development of their supra-œsophageal ganglion and sense-organs, and their powerful jaws and arms, or tentacles, show to what good purpose they have struggled. But the struggle was in vain, as far as the supremacy of the animal kingdom was concerned. Their ancestors had taken a course which rendered it impossible for their descendants to reach the goal. Their progress became ever slower. They were entirely and hopelessly beaten by the vertebrates. They struggled hard, but too late.

The history of mollusks is full of interest. They show clearly how intimately nervous development is connected with the use of the locomotive organs. The snail crept, and slightly increased its nervous system and sense-organs. The clam almost lost them in connection with its stationary life. The cephalopods were exceedingly active, developed, therefore, keen sense-organs and a very large and complicated supra-œsophagal ganglion, which we might almost call a brain.

The articulate series consists of two groups of animals. The higher group includes the crabs, spiders, thousand-legs, and finally the insects, and forms the kingdom of arthropoda. The lower members are still usually reckoned as worms, and are included under the annelids. Of these our common earthworm is a good example, and near them belong the leeches. But the marine annelids, of which nereis, or a clam-worm, is a good example, are more typical. They are often quite large, a foot or even more in length. They are composed of many, often several hundred, rings or segments. Between these the body-wall is thin, so that the segments move easily upon each other, and thus the animal can creep or writhe.

These segments are very much alike except the first two and the last. If we examine one from the middle of the body we shall find its structure very much like that of our schematic worm. Outside we find a very thin, horny cuticle, secreted by the layer of cells just beneath it, the hypodermis. Beneath the skin we find a thin layer of transverse muscles, and then four heavy

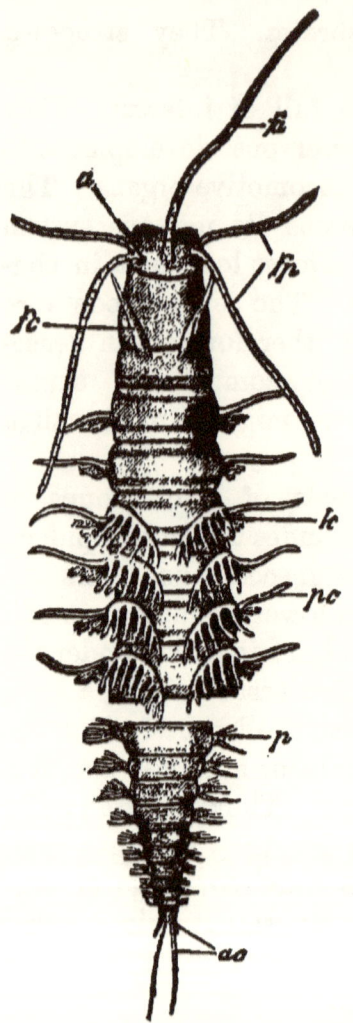

7. EUNICE LIMOSA (ANNELID).   LANG,
FROM EHLERS.

Front and hind end seen from dorsal
surface.

fa, fp, fc, feelers; *a,* eye; *k,* gill; *p,*
parapodia; *ac,* anal cirri.

bands of longitudinal muscles. These latter have been grouped in the four quadrants, a much more effective arrangement than the cylindrical layer of the schematic worm. Furthermore, the animal has on each segment a pair of fin-like projections, stiffened with bristles, the parapodia. These are moved by special muscles and form effective organs of creeping.

Within the muscles is the perivisceral cavity, and in its central axis the intestine, segmented like the body-wall. The reproductive organs are formed from patches of the lining of the perivisceral cavity, and the reproductive elements, when fully developed, fall into the perivisceral fluid and are carried out by nephridia, just such as we found in the schematic worm. Beside the perivisceral cavity and its fluid there is a special circulatory system. This consists mainly of one long tube above the intestine and a second below, with often several smaller parallel tubes. Transverse vessels run from these to all parts of the body. The dorsal tube pulsates and thus acts as a heart. The surface of the body no longer suffices to gather oxygen, hence we find special feathery gills on the parapodia. But these gills are merely expanded

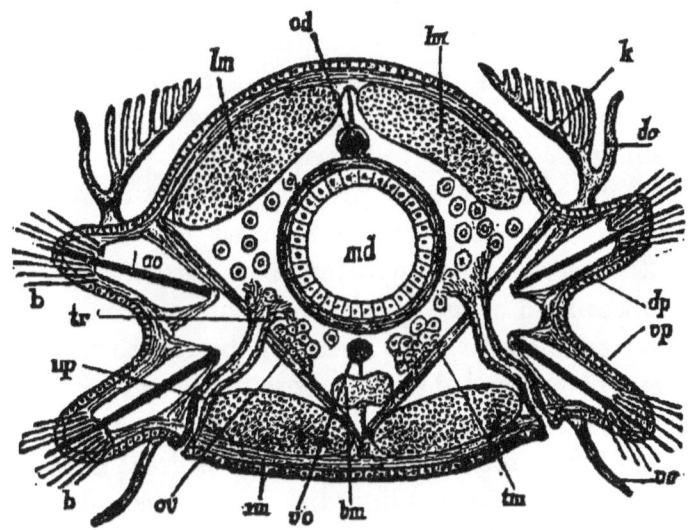

8. CROSS-SECTION OF BODY SEGMENT OF ANNELID. LANG.

dp and *vp*, dorsal and ventral halves of parapodia; *b* and *ac*, bristles; *k*, gill; *dc* and *vc*, feelers; *rm*, lateral muscles; *lm*, longitudinal muscles; *vd*, dorsal blood-vessel; *vo*, ventral blood-vessel; *bm*, ventral ganglion; *ov*, ovary; *tr*, opening of nephridium in the perivisceral cavity; *np*, tubular portion of nephridium. The circles containing dots represent eggs floating in the perivisceral fluid.

portions of the body wall, arranged so as to offer the greatest possible amount of surface where the capillaries of the blood system can be almost immediately in contact with the surrounding water.

The nervous system consists of a large supra-œsophageal ganglion in the first segment; then of a chain of ganglia, one to each segment, on the ventral side of the body. With one ganglion in each segment there is far more controlling, perceptive, ganglionic material than in lower worms. Furthermore the supra-œsophageal ganglion is relieved of a large part of the direct control of the muscles of each segment, and is becoming more a centre of control and perception for the body as a whole. It is more like our brain, commander-in-chief, the other ganglia constituting its staff. The sense-organs have improved greatly. There are tentacles and otolith vesicles as very delicate organs of feeling, or possibly of hearing also.

But the annelids were probably the first animals to develop an eye capable of forming an image of external objects. The importance of this organ in the pursuit of food or the escape from enemies can scarcely be over-estimated. The lining of the mouth and pharynx can be protruded as a proboscis, and drawn back by powerful muscles, and is armed with two or more horny claws. Eyes and claws gave them a great advantage over their not quite blind but really visionless and comparatively defenceless neighbors, and they must have wrought terrible extinction of lower and older forms. But while we cannot over-estimate the importance of these eyes, we can easily exaggerate their perfectness. They were of short range, fitted for seeing objects only a few inches distant, and the image was very imperfect in detail. But the plan or fundamental scheme of these eyes is correct and capable of indefinitely greater development than the organs of touch or smell, perhaps greater even than the otolith vesicle.

And the reflex influence of the eye on the brain was the greatest advantage of all. Hitherto with feeble muscles and sense-organs it has hardly paid the animal to devote more material to building a larger brain. It was better to build more muscle. But now with stronger muscles at its command, and better sense-organs to report to it, every grain of added brain material is beginning to be worth ten devoted to muscle. The muscular system will still continue to develop, but the brain has begun an almost endless march of progress. The eye becomes of continually increasing advantage and importance because it has a capable brain to use it; and brain is a more and more profitable investment, because it is served by an ever-improving eye.

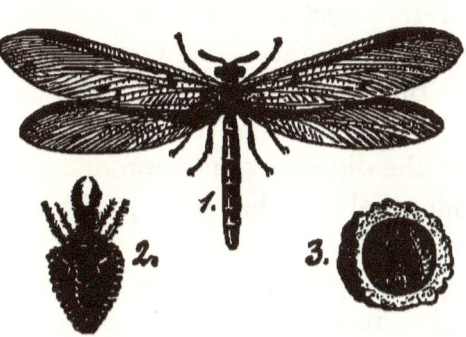

9. **MYRMELEO FORMICARIUS. ANT-LION. HERTWIG, FROM SCHMARDA.**

1, adult; 2, larva; 3, cocoon.

The annelid had hit upon a most advantageous line of development, which led ultimately to the insect. The study of the insect will show us clearly the advantages and defects of the annelid plan. First of all, the insect, like the mollusk, has an external skeleton. But the skeleton of the mollusk was purely protective, a hindrance to locomotion. That of the insect is still somewhat protective, but is mainly, almost purely, locomotive. It is never allowed to become so heavy as to interfere with locomotion. In the second place, the insect has three body regions, having each its own special functions or work. And one of these is a head. The annelid had two anterior segments differing from those of the rest of the body; these may,

5

perhaps, be considered as the foreshadowings of a structure not yet realized; they can only by courtesy be called a head. Thirdly, the insect has legs. The annelid had fin-like parapodia, approaching the legs of insects about as closely as the fins of a fish approach the legs of a mammal. The reproductive and digestive systems, while somewhat improved, are not very markedly higher than those of annelids. The excretory system has more work to perform and reaches a rather higher development.

But in these organs there is no great or striking change; the time for marked and rapid development of the digestive and reproductive systems has gone by. Material can be more profitably invested in brain or muscle. Air is carried to all parts of the body by a special system of air-sacks and tubes. This is a very advantageous structure for small animals with an external skeleton. In very large animals, or where the skeleton is internal, it would hardly be practicable; the risk of compression of the tubes at some point, and of thus cutting off the air-supply of some portion of the body, would be altogether too great.

The circulatory system is very poor. It consists practically only of a heart, which drives the blood in an irregular circulation between the other organs of the body much as with a syringe you might keep up a system of currents in a bowl of water. But the rapidity of the flow of the blood in our bodies is mainly to furnish a supply of oxygen to the organs. A teaspoonful of blood can carry a fair amount of dissolved solid nutriment like sugar, it can carry at each round but a very little gas like oxygen. Hence the blood must make its rounds rapidly, carrying but a little oxygen at each circuit. But in the insect the blood conveys only the dissolved solid nutriment, the food; hence a comparatively irregular circulation answers all purposes.

The skeleton is a thickening of the horny cuticle of the annelid on the surface of each segment. The horny cylinder surrounding each segment is composed of several pieces, and on the abdomen these are united by flexible, infolded membranes. This allows the increase in the size of the segment corresponding to the varying size of the digestive and reproductive systems. In this part of the body the skeletal ring of each segment is joined to that of the segments before and behind it in the same manner. But in other parts of the body we shall find the skeletal pieces of each segment and the rings of successive segments fused in one plate of mail. The legs are the parapodia of annelids carried to a vastly higher development. They are slender and jointed, and yet often very powerful. A large portion of the muscular system of the body is attached to these appendages.

But the insect has also jaws. The annelid had teeth or claws attached to the proboscis. But true jaws are something quite different. They always develop by modifying some other organ. In the insect they are modified legs. This is shown first by their embryonic development. But the king- or horseshoe-crab has still no true jaws, but uses the upper joints of its legs for chewing. There are primitively three pairs of jaws of various forms for the different kinds of food of different species or higher groups. But some of them may disappear and the others be greatly modified into awls for piercing, or a tube for sucking honey. Into the wonderful transformations of these modified legs we cannot enter.

The muscles are no longer arranged to form a sack as in annelids. Transverse muscles, running parallel to the unyielding plates of chitin or horn could accomplish nothing. They have largely disappeared. The work of locomotion has been transferred from the trunk to the legs.

The abdomen of the insect is as clearly composed of distinct segments as the body of the annelid. Of these there are perhaps typically eleven. The thorax is composed of three segments, distinct in the lowest forms, fused in the highest. This fusion of segments in the thorax of the highest forms furnishes a very firm framework for the attachment of wings and muscles. These wings are a new development, and how they arose is still a question. But they give the insect the capability of exceedingly rapid locomotion.

The three pairs of jaws, modified legs, in the rear half of the head show that this portion is composed of three segments. For only one pair of legs is ever developed on a single segment. Embryology has shown that the portion of the head in front of the mouth is also composed of three segments. Possibly between the præ- and post-oral portions still another segment should be included, making a total of seven in the head. The head has thus been formed by drawing forward segments from the trunk, and fusing them successively with the first or primitive head segment. This is difficult to conceive of in the fully developed insect, where the boundary between head and thorax is very sharp. But the ancestors of insects looked more like thousand-legs or centipedes, and here head and thorax are much less distinct. But in the annelid the mouth is on the second segment; here it is on the fourth. It has evidently travelled backward. That the mouth of an animal can migrate seems at first impossible, but if we had time to examine the embryology of annelids and insects, it would no longer appear inconceivable or improbable. And its backward migration brought it among the legs which were grasping and chewing the food. And in vertebrates the mouth has changed its position, though not in exactly the same way. Our present mouth is probably not at all the mouth of the primitive ancestor of vertebrates. Thus in the insect three segments have fused around the mouth, and three, possibly four, in front of it. This makes a head worthy of the name. The ganglia of the three post-oral segments, which bear the jaws, have fused in one compound ganglion innervating the mouth and jaws. Those of the three præ-oral segments have fused to form a brain. Eyes are well developed, giving images sometimes accurate in detail, sometimes very rude. Ears are not uncommon. The sense of smell is often keen.

Perhaps the greatest advance of the insect is its adaptation to land life. This gives it a larger supply of oxygen than any aquatic animal could ever obtain. This itself stimulates every function, and all the work of the body goes on more energetically. Then the heat produced is conducted off far less rapidly than in aquatic forms. Water is a good conductor of heat, and nearly all aquatic animals are cold-blooded. The few which are warm-blooded are protected by a thick layer of non-conducting fat. In all land animals, even when cold-blooded, the work of the different systems is aided by the longer retention of the heat in the body.

Let us recapitulate. The schematic worm had a body composed of two concentric tubes. The outer was composed of the muscles of the body covered by the protective integument. The inner tube was the alimentary canal with its special muscles. Between these two was the perivisceral cavity, filled with nutritive fluid, lymph, and furnishing a safe lodging-place for the more delicate viscera. It represented fairly the trunk of higher animals.

The annelid added segmentation, and thus greater freedom of motion by the parapodia. But the segments were still practically alike. In the insect division of labor took place, that is, each group of segments was allotted its own special work; and these groups of segments were modified in structure to best suit the performance of this part of the work of the body. The abdomen was least modified and its eleven segments were devoted to digestion, reproduction, and excretion—the old vegetative functions. Three segments were united in the thorax; all their energy was turned to locomotion, and the insect became thus an exceedingly active, swift animal. The third body-region, the head, includes six segments, of which three surrounded the mouth and furnished the jaws, while two more were crowded or drawn forward in order that their ganglia might be added to the old supracœsophageal ganglion and form a brain. It is interesting to note that a form, peripatus, still exists which stands almost midway between annelids and insects and has only four segments in the head. The formation of the head was thus a gradual process, one segment being added after another.

In the turbellaria the dominant functions were digestion and reproduction, and their organs composed almost the whole body. Here only eleven segments at most are devoted to these functions, and nine in head and thorax to locomotion and brain. Head and thorax have increased steadily in importance, while the abdomen has decreased as steadily in number of segments. And the brain is increasing thus rapidly because there are now muscles and sense-organs of sufficient power to make such a brain of value. And this brain perceives not only objects and qualities, but invisible relations between these, and this is an advance amounting to a revolution. It remembers, and uses its recollections. It is capable of learning a little by experience and observation. The A, B, C of thinking was probably learned long before the insect's time, and the bee shows a fair amount of intelligence.

The line of development which the insect followed was comparatively easy and its course probably rapid. Certain crustacea, aquatic arthropoda, are among the oldest fossils, and it is possible that insects lived on the land before the first fish swam in the sea. They had fine structure and powers; and yet during the later geologic periods they have scarcely advanced a step, and are now apparently at a standstill. They ran splendidly for a time, and then fell out of the race. What hindered and stopped them?

One vital defect in their whole plan of organization is evident. The external skeleton is admirably suited to animals of small size, but only to these. In larger animals living on land it would have to be made so heavy as to be unwieldy and no longer economical. Their mode of breathing also is fitted only for animals of small size having an external skeleton. Whatever may be our explanation the fact remains that insects are always small. This is in itself a disadvantage. Very small animals cannot keep up a constant high temperature unless the surrounding air is warm, for their radiating surface is too large in comparison with their heat-producing mass. At the first approach of even cool weather they become chilled and sluggish, and must hibernate or die. They are conformed to but a limited range of environment in temperature.

But small size is, as a rule, accompanied by an even greater disadvantage. It seems to be almost always correlated with short life. Why this is so, or how, we do not know. There are exceptions; a crow lives as long as a man ; or would, if allowed to. But, as a rule, the length of an animal's days is roughly proportional to the size of its body. And the insect is, as a rule, very short-lived. It lives for a few days or weeks, or even months, but rarely outlasts the year. It has time to learn but little by experience. The same experience must be passed, the same emergency arise and be met, over and over again during the lifetime of the same individual if the animal is to learn thereby. And intelligence is based upon experience. Hence insects can and do possess but a low grade of intelligence. But instinct is in many cases habit fixed by heredity and improved by selection. The rapid recurrence of successive generations was exceedingly favorable to the development of instincts, but very unfavorable to intelligence. Insects are instinctive, the highest vertebrates intelligent. The future can never belong to a tiny animal governed by instincts. Mollusks and insects have both failed to reach the goal; another plan of structure than theirs must be sought if the animal kingdom is to have a future.

The future belonged to the vertebrate. To begin with less characteristic organs the digestive system is much like that of the annelid or schematic worm, but with greatly increased glandular and absorptive surfaces. The present mouth of nearly all vertebrates is probably not primitive. It is almost certainly one of the gill-slits of some old ancestor of fish, such as now are used to discharge the water which is used for respiration. The jaws are modified branchial arches or the cartilaginous or bony rods which in our present fish support the fringe of gills. These have formed a pair of exceedingly effective and powerful jaws. The reproductive system holds still to the old type and shows little if any improvement. The excretory organs, kidneys, are composed primitively of nephridial tubes like those of the schematic worm or annelid, but immensely increased in number, modified, and improved in certain very important particulars. The muscles in simplest forms are composed of heavy longitudinal bands, especially developed toward the dorsal surface of the body to the right and left of the axial skeleton. Locomotion was produced by lashing the tail right and left, as still in fish. There is improvement in all these organs, except perhaps the reproductive, but nothing very new or striking. The great improvement from this time on was not to be sought in the vegetative organs, or even directly to any great extent in muscles.

The new and characteristic organ was not the vertebral column, or series of vertebræ, or backbone, from which the kingdom has derived its name. This was a later production. The primitive skeleton was the notochord, still appearing in the embryos of all vertebrates and persisting throughout life in fish. This is an elastic rod of cartilage, lying just beneath the spinal marrow or nerve-cord, which runs backward from the brain. The nerve-centres are therefore here all dorsal, and the notochord or skeleton lies between these and the digestive or alimentary canal. The skeleton of the clam or snail is purely protective and a hindrance to locomotion. That of the insect is almost purely locomotive, but external, that of the vertebrate purely locomotive and internal. It does not lie outside even of the nervous system, although this system especially required, and was worthy of, protection. It does not protect even the brain; the skull of vertebrates is an after-thought. It is almost the deepest seated of all organs.

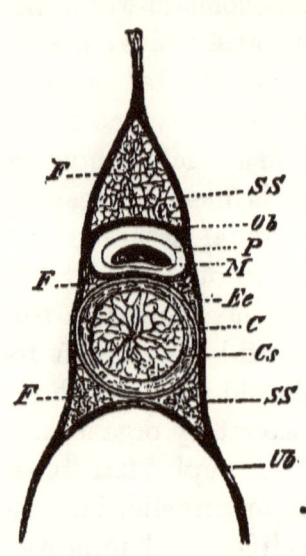

10. CROSS-SECTION OF AXIAL SKELETON OF PETROMYZON. HERTWIG, FROM HIEDERSHEIM.

SS, skeletogenous layer; *Ob*, *Ub*, dorsal and ventral processes of *SS*; *C*, notochord; *Cs*, sheath of notochord; *Ee*, elastic external layer of sheath; *F*, fatty tissue; *M*, spinal marrow; *P*, sheath of *M*.

But lying in the central axis of the body it furnishes the very best possible attachment for muscles. Around this primitive notochord was a layer of connectile tissue which later gave rise to the vertebræ forming our backbone.

The nervous system on the dorsal surface of the notochord consists of the brain in the head and the spinal marrow running down the back. The brain of all except the very lowest vertebrates consists of four portions : 1. The cerebrum, or cerebral lobes, or simply "forebrain," the seat of consciousness, thought, and will, and from which no nerves proceed. Whether the primitive vertebrate had any cerebrum is still uncertain. 2. The mid-brain, which sends nerves to the eyes, and in this respect reminds us of the brain of insects. Its anterior portion appears from embryology to be very primitive. 3. The small brain, or cerebellum, which in all higher forms is the centre for co-ordination of the motions of the body. 4. The medulla, which controls especially the internal organs. The spinal marrow, or that portion of the nervous system which lies outside of the head, is at the same time a great nerve-trunk and a centre for reflex action of the muscles of the body. But the development of these distinct portions and the division of labor between them must have been a long and gradual process.

We have every reason to believe that here, as in insects, the head has been formed by annexation of segments from the rump and the fusion of their nervous matter with that of the brain. But here, instead of only three segments, from nine to fourteen have been fused in the head to furnish the material for the brain. Notochord and backbone may be the most striking and apparent characteristic of vertebrates, but their predominant characteristic is brain. On this system they lavished material, giving it from three to four times as much as any lower or earlier group had done. They very early set apart the cerebral lobes to be the commander-in-chief and centre of control for all other nerve-centres. To this all report, and from it all di-

rectly or indirectly receive orders.   It can say to every
other organ in the body, "Starve that I may live."   It
is the seat of thought and will.   The other portions of
the brain report to it what they have gathered of vision
or sound; it explains the vision or song or parable.
It is relieved as far as possible from all lower and
routine work that it may think and remember and
govern.   The vertebrate built for mind, not neglecting
the body.

Every trait of vertebrates is a promise of a great future.   Its internal skeleton gives it the possibility of
large size.   This gave it in time the victory in the
struggle with its competitors, as to whether it should
eat or be eaten.   It is vigorous and powerful, for all its
organs are at the best.   It gives the possibility of later, on land, becoming warm-blooded, *i.e.*, of maintaining a constant high temperature.   It is thus resistant
to climate and hardship.   In time its descendants will
face the arctic winter as well as the heat of the tropics.

But it has started on the road which leads to mind.
The greater size is correlated with longer life.   The
lessons of experience come to it over and over again,
and it can and must learn them.   It is the intelligent,
remembering, thinking type.   The insect had begun to
peer into the world of invisible and intangible relations,
the vertebrate will some day see them.   This much is
prophecied in his very structure.   He must be heir to
an indefinite future.

You have probably noticed that the vertebrate differs greatly from all his predecessors.   The gulf
between him and them is indeed wide and deep.   His
origin and ancestry are yet far from certain.   But an

attempt to decipher his past history, though it may lead to no sure conclusions, will yet be of use to us. Practically all aquatic vertebrates lead a swimming life, neither sessile nor creeping. The embryonic development of our appendages leads to the same conclusion. We must never forget that the embryonic development of the individual recapitulates briefly the history of the development of the race. Now the legs and arms, or fore- and hind-legs, of higher vertebrates and the corresponding paired fins of fish develop in the embryo as portions of a long ridge extending from front to rear of the side of the body.

This justifies the inference that the primitive vertebrate ancestor had a pair of long fins running along the sides of the body, but bending slightly downward toward the rear so as to meet one another and continue as a single caudal fin behind the anal opening. Such fins, like the feathers of an arrow, could be useful only to keep the animal " on an even keel " as it was forced through the water by the lateral sweeps of the tail. They would have been useless for creeping.

But there is another piece of evidence that he was a free swimming form. All vertebrates breathe by gills or lungs, and these are modified portions of the digestive system, of the walls of the œsophagus, from which even the lung is an embryonic outgrowth. Now practically all invertebrates breathe through modified portions of the integument or outer surface of the body, and their gills are merely expansions of this. In the annelid they are projections of the parapodia, in the mollusk expansions of the skin, where the foot or creeping sole joins the body. Why did the vertebrate take a new and strange, and, at first sight, disadvantageous mode of breathing? There must have been some good reason for this. The most natural explanation would seem to be that he had no projections on his outer surface which could develop into gills, and farther, that he could not afford to have any. Now projections on the lower portion of the sides of the body would be an advantage in creeping, but a hindrance in any such mode of swimming as we have described, or indeed in any mode of writhing through the water.

Furthermore, if he lived, not a creeping life on the bottom, but swimming in the water above, he would have to live almost entirely on microscopic animals and embryos; and these would be most easily captured by a current of water brought in at the mouth. The whole branchial apparatus in its simplest forms would seem to be an apparatus for sifting out the microscopic particles of food and only later a purely respiratory apparatus. Moreover, we have seen that the parapodia of annelids naturally point to the development of an external skeleton, for their muscles are already a part of the external body-wall and attached to the already existing horny cuticle. The logical goal of their development was the insect.

Now I do not wish to conceal from you that many good zoölogists believe that the vertebrate is descended from annelids; but for this and other reasons such a descent appears to me very improbable. It would seem far more natural to derive the vertebrate from some free swimming form like the schematic worm, whose largest nerve-cord lay on the dorsal surface because its branches ran to heavy muscles much used in swimming. Later the other nerve-cords degenerated, for such a degeneration of nerve-cords is not at all impossible or improbable. "No thoroughfare" is often written across paths previously followed by blood or nervous impulses, when other paths have been found more economical or effective.

But where did the notochord come from? I do not know. It always forms in the embryo out of the entoderm or layer which becomes the lining of the intestine. Now this is a very peculiar origin for cartilage, and the notochord is a very strange cartilage even if we have not made a mistake in calling it cartilage at all. My best guess would be that it is simply a thickened portion of the upper median surface of the intestine to keep the "balls" of digesting nutriment or other hard particles in the intestine from "grinding" against the nerve-cord as they are crowded along in the process of digestion. Once started its elasticity would be a great aid in swimming.

Professor Brooks has called attention to the fact that the higher a group stands in development, the longer its ancestors have maintained a swimming life. Thus we have noticed that the sponges were the first to settle; then a little later the mass of the cœlenterates followed their example. But the ctenophora, the nearest relatives of bilateral animals, have remained free swimming. Then the flat worms and mollusks took to a creeping mode of life, while the annelids and vertebrates still swam. Then the annelids settled to the bottom and crept, and all their descendants remained creeping forms. The vertebrates alone remained swimming, and probably neither they nor their descendants ever crept until they emerged on the land, or as amphibia were preparing for land life. If this be true, it is a fact worthy of our most careful consideration. The swimming life would appear to be neither as easy nor as economical as the creeping. It is certainly hard to believe that food would not have been obtained with less effort and in greater abundance at the bottom than in the water above. The swimming life gave rise to higher and stronger forms; but did its maintenance give immediate advantage in the struggle for existence? This is an exceedingly interesting and important question, and demands most careful consideration. But we shall be better prepared to answer it in a future lecture.

The period of development of mollusks, articulates, and vertebrates, is really one. They developed to a certain extent contemporaneously. The development of vertebrates was slow, and they were the last to appear on the stage of geological history.

You must all have noticed that development, during this period, takes on a much more hopeful form than during that described in the last chapter. Then digestion and reproduction were dominant. Now muscle is of the greatest importance. If this fails of development, as in mollusks, the group is doomed to degeneration or at best stagnation. But we have seen the dawn of a still higher function. In insects and vertebrates the brain is becoming of importance, and absorbing more and more material. This is the promise of something vastly higher and better. Better sense-organs are appearing, fitted to aid in a wider perception of more distant objects. The vertebrate has discovered the right path; though a long journey still lies before it. The night is far spent, the day is at hand.

CHAPTER IV

In tracing man's ancestry from fish upward we ought properly to describe three or four fish, an amphibian, a reptile, and then take up the series of mammalian ancestors. But we have not sufficient time for so extended a study, and a simpler method may answer our purpose fairly well. Let us fix our attention on the few organs which still show the capacity of marked development, and follow each one of these rapidly in its upward course.

We must remember that there are changes in the vegetative organs. The digestive and excretory systems improve. But this improvement is not for the sake of these vegetative functions. Brain and muscle demand vastly more fuel, and produce vastly more waste which must be removed. At almost the close of the series the reproductive system undergoes a modification which is almost revolutionary in its results. But we shall find that this modification is necessitated by the smaller amount of material which can be spared for this function ; not by its increasing importance, still less its dominance for its own worth. The vertebrate is like an old Roman ; everything is subordinated to mental and physical power. He is the world conqueror.

6

The important changes from fish upward affect the following organs : 1. The skeleton. A light, solid framework must be developed for the body. 2. The appendages start as fins, and end as the legs and arms of man. 3. The circulatory and respiratory systems developed so as to carry with the utmost rapidity and certainty fuel and oxygen to the muscular and nervous high-pressure engines. Or, to change the figure, they are the roads along which supplies and munitions can be carried to the army suddenly mobilized at any point on the frontier. 4. Above all, the brain, especially the cerebrum, the crown and goal of vertebrate structure. The improvement is now practically altogether in the animal organs of locomotion and thought. Still, among these animal organs, the lower systems will lead in point of time. The brain must to a certain extent wait for the skeleton.

1. The skeleton. The axial skeleton consists, in the lowest fish, of the notochord, a cylindrical unsegmented rod of cartilage running nearly the length of the body. This is surrounded by a sheath of connective tissue, at first merely membranous, later becoming cartilaginous or gristly. Pieces of cartilage extend upward over the spinal marrow, and downward around the great aortic artery, forming the neural and hæmal arches. These unite with the masses of cartilage surrounding the notochord to form cartilaginous vertebræ, which may be stiffened by an infiltration of carbonate of lime. The vertebral column of sharks has reached this stage. Then the cartilaginous vertebræ ossify and form a true backbone. I have described the process as if it were very simple. But only the student of comparative osteology can have any con-

ception of the number of experiments which were tried in different groups before the definite mode of forming a bony vertebra was attained. At the same time the skull was developing in a somewhat similar manner. But the skull is far more complex in origin and undergoes far more numerous and important changes than the simpler vertebral column. Into its history we have no time to enter.

And what shall we say of bone itself as a mere material or tissue, with its admirable lightness, compactness, and flawlessness. And every bone in our body is a triumph of engineering architecture. No engineer could better recognize the direction of strain and stress, and arrange his rods and columns, arches and buttresses, to suitably meet them, than these problems are solved in the long bone of our thigh. And they must be lengthened while the child is leaping upon them. An engineer is justly proud if he can rebuild or lengthen a bridge without delaying the passage of a single train. But what would he say if you asked him to rebuild a locomotive, while it was running even twenty miles an hour? And yet a similar problem had to be solved in our bodies.

But the vertebral column is not perfected by fish. The vertebræ with few exceptions are hollow in front and behind, biconcave; and between each two vertebræ there is a large cavity still occupied by the notochord. Thus these vertebræ join one another by their edges, like two shallow wine-glasses placed rim to rim. Only gradually is the notochord crowded out so that the vertebræ join by their whole adjacent surfaces. Even in highest forms, for the sake of mobility, they are united by washer-like disks of cartilage. Biconcave vertebræ persisted through the oldest amphibia, reptiles, and birds. But finally a firm backbone and skull were attained.

2. The appendages. Of these we can say but little. The fish has oar-like fins, attached to the body by a joint, but themselves unjointed. By the amphibia legs, with the same regions as our own and with five toes, have already appeared. The development of the leg out of the fin is one of the most difficult and least understood problems of vertebrate comparative anatomy. The legs are at first weak and scarcely capable of supporting the body. Only gradually do they strengthen into the fore- and hind-legs of mammals, or into the legs and wings of birds and old flying reptiles.

3. Changes in the circulatory and respiratory systems. The fish lives altogether in the water and breathes by gills, but the dipnoi among fishes breathes by lungs as well as gills. As long as respiration takes place by gills alone, the circulation is simple ; the blood flows from the heart to the gills, and thence directly all over the body ; the oxygenated blood from the gills does not return directly to the heart. But the blood from the lungs does return to the heart ; and there at first mixes in the ventricle with the impure blood which has returned from the rest of the body. Gradually a partition arises in the ventricle, dividing it into a right and left half. Thus the two circulations of the venous blood to the lungs, and of the oxygenated blood over the body, are more and more separated until, in higher reptiles, they become entirely distinct.

As the animal came on land and breathed the air, more completely oxygenated blood was carried to the organs, and their activity was greatly heightened. As more and more heat was produced by the combustion in muscular and nervous tissues, and less was lost by conduction, the temperature of the body rose, and in birds and mammals becomes constant several degrees above the highest summer temperature of the surrounding air.

The changes in the brain affect mainly the large and small brain. The cerebellum increases with the greater locomotive powers of the animal. But its development is evidently limited. The large brain, or cerebrum, is in fish hardly as heavy as the mid-brain; in amphibia the reverse is true. In higher recent reptiles the cerebrum would somewhat outweigh all the other portions of the brain put together. In mammals it extends upward and backward, has already in lower forms overspread the mid-brain, and is beginning to cover the small brain. But this was not so in the earliest mammals. Here the cerebrum was small, more like that of reptiles. But during the tertiary period the large brain began to increase with marvellous rapidity. It was very late in arriving at the period of rapid development, but it kept on after all the other organs of the body had settled down into comparative rest, perhaps retrogression.

We have given thus a rapid sketch in outline of the changes in the most characteristic systems between fish and mammals. Some of the changes which took place in mammals were along the same lines, but one at least is so new and unexpected that this highest class demands more careful and detailed examination.

The mammal is a vertebrate. Hence all its organs are at their best. But mammals stand, all things considered, at the head of vertebrates. The skeleton is firm and compact. The muscles are beautifully moulded and fitted to the skeleton so as to produce the greatest effect with the least mass and weight of tissue. The sense-organs are keen, and the eye and ear especially delicate, and fitted for perception at long range. Yet in all these respects they are surpassed by birds. As a mere anatomical machine the bird always seems to me superior to the mammal. It is not easy to see why it failed, as it has, to reach the goal of possibility of indefinite development and dominance in the animal world. Why he stopped short of the higher brain development I cannot tell. The fact remains that the mammal is pre-eminent in brain power, and that this gave him the supremacy.

But mammals came very late to the throne, and the probability of their ever gaining it must for ages have appeared very doubtful. They seem to have been a fairly old group with a very slow early development. Reptiles especially, and even birds, were far more precocious than these slower and weaker forms which crept along the earth. But reptiles and birds, like many other precocious children, soon reached the limit of their development. They had muscle, the mammal brain and nerve ; the mammal had the staying power and the future. Bitter and discouraging must have been the struggle of these feeble early mammals with their larger, swifter, and more powerful, reptilian relatives. And yet, perhaps, by this very struggle the mammal was trained to shrewdness and endurance.

The primitive mammals laid eggs like reptiles or birds. Only two genera, echidna and platypus, sur-

vive to bear witness of these old oviparous groups, and
these only in New Zealand. These retain several old
reptilian characteristics. Their lower position is shown
also by the fact that the temperature of their bodies is,
at least, ten degrees Fahrenheit below that of higher
mammals. One of these carries the egg in a pouch on
the ventral surface; the other, living largely in water,
deposits its eggs in a nest in a burrow in the side of
the bank of the stream.

After these came the marsupials. In these the eggs
develop in a sort of uterns; but there is no placenta,
in the sense of an organic connection between the embryo and the uterus of the mother. The young are at
birth exceedingly small and feeble. The adult giant
Kangaroo weighs over one hundred pounds; the young
are at birth not as large as your thumb. They are
placed by the mother in a marsupial pouch on her
ventral surface, and here nourished till able to care for
themselves.

Pardon a moment's digression. The marsupials, except the opossum, are confined to Australia, and the
oviparous mammals, or monotremes, to New Zealand.
Formerly the marsupials, at least, ranged all over
Europe and Asia, for we have indisputable evidence in
their fossil remains. But they have survived only in
this isolated area, and here apparently only because
their isolation preserved them from the competition
with higher forms. If the Australian continent had
not been thus early cut off from all the rest of the
world, the only trace of both these lower groups would
have been the opossum in America and certain peculiarities in the development of the egg in higher
mammals. This shows us how much weight should be

assigned to the formerly popular argument of the "missing links." The wonder is not that so many links are missing, but that any of these primitive forms have come down to us. For we see here another proof of the fearful extermination of lower forms during the progress of life on the globe. It seems as if the intermediate forms were less common among these most recent animals than among the older types. This may not be true, for it is not easy to compare the gap between two mammals with that between two worms or insects, and mistakes are very easily made. But it seems as if extermination had done its work more ruthlessly among these highest forms than among their humbler and lower ancestors. I would not lay much weight on such an opinion; but, if true, it has a meaning and is worthy of study.

In higher, true, placental mammals the period of pregnancy is much longer, and the young are born in a far higher stage of development, or rather, growth. The stage of growth at which the young are born differs markedly in different groups. A new-born kitten is a much feebler, less developed being than a new-born calf. An embryonic appendage, the allantois, used in reptiles and birds for respiration, has here been turned to another purpose. It lays itself against the walls of the uterus, uterine projections interlock with those which it puts forth, and the blood of the mother circulates through a host of capillaries separated from those of the blood system of the embryo only by the thinnest membrane. This is the placenta, developed, in part from the allantois of the embryo, in part from the uterus of the mother. It is not a new organ, but an old one turned to better and fuller use.

In these closely associated systems of blood-vessels, nutriment and oxygen diffuse from the blood of the mother into that of the embryo, and thus rapid growth is assured. The importance and far-reaching effect of this new modification in the old reproductive system cannot be over-estimated. The internal intra-uterine development of the young, and the mammalian habit of suckling them, far more than any other factors, have made man what he is. Some explanation must be sought for such a fact.

We have already seen that any animal devotes to reproduction the balance between income and expenditure of nutriment. Now, the digestive system is here well developed, and the income is large. But we have already noticed that, as animals grow larger, the ratio between the digestive surface and the mass to be supported grows continually smaller. On account of size alone the mammal has but a small balance. But the amount of expenditure is proportional to the mass and activity of the muscular and nervous systems. And the mammal is, and from the beginning had to be, an exceedingly active, energetic, and nervous animal. The income has increased, but the expenses have far outrun the increase. The mammal can devote but little to reproduction.

Moreover, it requires a large amount of material to form a mammalian egg, such as that of the monotreme. It requires indefinitely more nutriment to build a mammal than a worm, for the former is not only larger and more perfect at birth; it is also vastly more complicated. The embryonic journey has, so to speak, lengthened out immensely. One monotreme egg represents more economy and saving than a thousand eggs of a worm. Moreover, where the individuals are longer lived and the generations follow one another at longer intervals, the number of favorable variations and the possibility of conformity to environment through these is greatly lessened. In such a group it is of the utmost importance that every egg should develop; the destruction of a single one is a real and important loss to the species. It is not enough to produce such an egg; it must be most scrupulously guarded. Even the egg of the platypus is deposited in a nest in a hole in the bank, and the female Echidna carries the egg in a marsupial pouch until it develops.

Notice further that among certain species of fish, amphibia, and reptiles, the females carry the eggs in the body until the embryos or young are fairly developed. Viviparous forms are unknown by birds, probably because this mode of development is incompatible with flight, their dominant characteristic. Putting these facts together, what more probable than that certain primitive egg-laying mammals should have carried the eggs as long as possible in the uterus. The embryo under these conditions would be better nourished by a secretion of the uterine glands than by a very large amount of yolk. The yolk would diminish and the egg decrease in size, and thus the marsupial mode of development would have resulted. And, given the marsupial mode of development and an embryo possessing an allantois, it is almost a physiological necessity that in some forms at least a placenta should develop. That the placenta has resulted from some such process of evolution is proven by its different stages of development in different orders of mammals. And even the feeblest attachment of the allantois of the embryo to the wall of the uterus would be of the greatest advantage to the species.

This is not the whole explanation; other factors still undiscovered were undoubtedly concerned. But even this shows us that the internal development of the young and the habit of suckling them was a logical result of mammalian structure and position. The grand results of this change we shall trace farther on.

The changes from the lower true mammals to the apes are of great interest, but we can notice only one or two of the more important. The prosimii, or "half apes," including the lemurs, are nearly all arboreal forms. Perhaps they were driven to this life by their more powerful competitors. The arboreal life developed the fingers and toes, and most of these end, not with a claw, but with a nail. The little group has much diversity of structure, and at present finds its home mainly in Madagascar; though in earlier times apparently occurring all over the globe. The brain is more highly developed than in the average mammal, but far inferior to that of the apes. They have a fairly opposable thumb.

The highest mammals are the primates. Their characteristics are the following: Fingers and toes all armed with nails, the eyes comparatively near together and fully enclosed in a bony case. The cerebrum with well-developed furrows covers the other portions of the brain. There is but one pair of milk-glands, and these on the breast. The differences between hand and foot become most strongly marked by the "anthropoid" apes. These have become accustomed to an upright gait in their climbing; hence the feet are used for supporting the body and the hands for grasping. Both thumb and great toe are opposable ; but the foot is a true foot, and the hand a true hand, in anatomical structure. The face, hands, and feet have mainly lost the covering of hair. They have no tail, or rather its rudiments are concealed beneath the skin. These include the gibbon, the orang, the gorilla, and the chimpanzee.

We can sum up the few attainments of mammals in a line. The lower forms attained the placental mode of embryonic development; the higher attained upright gait, hands and feet, and a great increase of brain. Anatomically considered these were but trifles, but the addition of these trifles revolutionized life on the globe. The principal anatomical differences between man and the anthropoid ape are the following: Man is a strictly erect animal. The foot of the ape is less fitted for walking on the ground, where he usually "goes on all fours." The skull is almost balanced on the condyles by which it articulates with the neck, and has but slight tendency to tip forward. The facial portion, nose and jaws, is less developed and retracted beneath the larger cranium or brain-case. This has greatly changed the appearance of the head. Protruding jaws and chin, even when combined with large cranium and brain, always give man the appearance of brutality and low intelligence.

The pelvis is broad and comparatively shallow. The legs, especially the thighs, are long. The foot is long and strong, and rests its lower surface, not merely the outer margin as in apes, on the ground. The elastic arch of the instep must be excepted in the above description, and adds lightness and swiftness to his otherwise slow gait. The great toe is short and generally not opposable. The muscles of the leg are heavy and the knee-joint has a very broad articulating surface. But the great result of man's erect posture is that the hand is set free from the work of locomotion, and has become a delicate tactile and tool-using organ. The importance of this change we cannot over-estimate. The hand was the servant of the brain for trying all experiments. Had not our arboreal ancestors developed the hand for us we could never have invented tools nor used them if invented. And its reflex influence in developing the brain has been enormous. The arm is shorter and the hand smaller. The brain is absolutely and relatively large, and its surface greatly convoluted. This gives place for a large amount of "gray matter," whose functions are perception, thought, and will. For this gray matter forms a layer on the outside of the brain.

Thus, even anatomically, man differs from the anthropoid apes. His whole structure is moulded to and by the higher mental powers, so that he is the "Anthropos" of the old Greek philosophers, the being who "turns his face upward." Yet in all these anatomical respects some of the apes differ less from him than from the lower apes or "half apes." And every one of these can easily be explained as the result of progressive development and modification. Whoever will deny the possibility or probability of man's development from some lower form must argue on psychological, not on anatomical, grounds ; and it grows clearer every day that even the former but poorly justify such a denial.

But it is interesting to note that no one ape most closely approaches man in all anatomical respects.

Thus among the anthropoids the orang is perhaps most similar to man in cerebral structure, the chimpanzee in form of skull, the gorilla in feet and hands. No evolutionist would claim that any existing ape represents the ancestor of man. The anthropoids represent very probably the culmination of at least three distinct lines of development. But we must remember that in early tertiary times apes occurred all over Europe, and probably Asia, many degrees farther north than now. In those days, as later, the fauna and flora of northern climates were superior in vigor and height of development to that of Africa or Australia. It is thus, to say the least, not at all improbable that there existed in those times apes considerably, if not far, superior to any surviving forms. Whether the palæontologist will find for us remains of such anthropoids is still to be seen.

But you will naturally ask, " Is there not, after all, a vast difference between the brain of man and that of the ape?" Let us examine this question as fully as our very brief time will allow. Considerable emphasis used to be laid on the facial angle between a line drawn parallel to the base of the skull and one obliquely vertical touching the teeth and most prominent portion of the forehead. Now this angle is in man very large—from seventy-five to eighty-five degrees, or even more, and rarely falling below sixty-five degrees. But this angle depends largely on the protrusion of the jaws, and varies greatly in species of animals showing much the same grade of intelligence. In some not especially intelligent South American monkeys the facial angle amounts to about sixty-five degrees. In this respect the skull of a chimpanzee reminds us of a human skull of small cranial capacity and large jaws, in which the cranium has been pressed back and the jaws crowded forward and slightly upward.

The weight of the brain in proportion to that of the body has been considered as of great importance, and within certain limits this is undoubtedly correct. Thus, according to Leuret, the weight of the brain is to that of the whole body : In fish, 1 : 5,668 ; in reptiles, 1 : 1,320 ; in birds, 1 : 212 ; in mammals, 1 : 186. These figures give the averages of large numbers of observations and have a certain amount of value. But within the same class the ratio varies extraordinarily. Thus the weight of the brain is to that of the whole body : In the elephant, 1 : 500; in the largest dogs, 1 : 305; in the cat, 1 : 156; in the rat, 1 : 76 ; in the chimpanzee, 1 : 50; in man, 1 : 36 ; in the field-mouse, 1 : 31; in the goldfinch, 1 : 24.

From this series it is evident that the relative weight of the brain is no index of the intelligence of the animal. Indeed if the brain were purely an organ of mind, there is no reason that it should be any larger in an elephant than in a mouse, provided they had the same mental capacity. As animals grow larger the weight of the brain, relatively to that of the body, decreases, and considering the size of man it is remarkable that it should form so large a fraction of his weight. Still the fraction in the chimpanzee is not so much smaller. It is still possible that this fraction is above the normal for the chimpanzee, for some of the observations may have been taken on animals which had died of consumption or some other wasting disease. I have not been able to find whether this possibility of error has been scrupulously avoided.

A fair idea of the size of the brain may be obtained by measuring the cranial capacity. This varies in man from almost one-hundred cubic inches to less than seventy. In the gorilla its average is perhaps thirty, in the orang and chimpanzee rather less, about twenty-eight. This is certainly a vast difference, especially when we remember that the gorilla far exceeds man in weight.

Le Bon tells us that of a series of skulls forty-five per cent. of the Australian had a cranial capacity of 1,200 to 1,300 c.c, while 46.7 per cent. of modern Parisian skulls showed a capacity of between 1,500 and 1,600 c.c. The skull of the gorilla contains about five hundred and seventy cubic centimetres. Broca found that the cranial capacity of 115 Parisian skulls, of probably the higher classes from the twelfth century, averaged about 1,426 cubic centimetres, while ninety of those of the poorer classes of the nineteenth century averaged about 1,484. His observations seemed to prove that there has been a steady increase in Parisian cranial capacity from the twelfth to the nineteenth century.

Turning to the actual weight of the brain, that of Cuvier weighed 64.5 ounces, and a few cases of weights exceeding 65 ounces have been recorded. The lowest limit of weight in a normal human brain has not yet been accurately determined. From 34 to 31 ounces have been assigned by different writers. The brain of a Bushwoman was computed by Marshall at 31.5 ounces, and weights of even 31 ounces have been recorded without any note to show that the possessors were especially lacking in intelligence. As Professor Huxley says in his "Man's Place in Nature," a little book which I cannot too highly recommend to you all, "It may be doubted whether a healthy human adult brain ever weighed less than 31 or 32 ounces, or that the heaviest gorilla brain has ever exceeded 20 ounces. The difference in weight of brain between the highest and the lowest men is far greater, both relatively and absolutely, than that between the lowest man and the highest ape. The latter, as has been seen, is represented by 12 ounces of cerebral substance absolutely, or by 32 : 20 relatively. But as the largest recorded human brain weighed between 65 and 66 ounces, the former difference is represented by 33 ounces absolutely, or by 65 : 32 relatively."

But there is another characteristic of the brain which seems to bear a close relation to the degree of intelligence. The surface of the human brain is not smooth but covered with convolutions, with alternating grooves or sulci, which vastly increase its surface and thus make room for more gray matter. Says Gratiolett: "On comparing a series of human and simian brains we are immediately struck with the analogy exhibited in the cerebral forms in all these creatures. There is a cerebral form peculiar to man and the apes; and so in the cerebral convolutions, wherever they appear, there is a general unity of arrangement, a plan, the type of which is common to all these creatures." Professor Huxley says : "It is most remarkable that, as soon as all the principal sulci appear, the pattern according to which they are arranged is identical with the corresponding sulci in man. The surface of the brain of the monkey exhibits a sort of skeleton map of man's, and in the man-like apes the details become more and more filled in, until it is only in minor

7

characters that the chimpanzee's or orang's brain can be structurally distinguished from man's.

The facts of anatomy, at least, are all against us. Struggle as we may, be as snobbish as we will, we cannot shake off these poor relations of ours. Our adult anatomy at once betrays our ancestry, if we attempt to deny it. Read the first chapter of that remarkable book by Professor Drummond on the "Ascent of Man," the chapter on the ascent of the body, and the second chapter on the scaffolding left in the body. The tips of our ears and our rudimentary ear muscles, the hair on hand and arm, and the little plica semilunaris, or rudimentary third eyelid in the inner angle of our eyes, the vermiform appendage of the intestine, the coracoid process on our shoulder-blades, the atlas vertebra of our necks—to say nothing of the coccyx at the other end of the backbone—many malformations, and a host of minor characteristics all refute our denial.

If we appeal from adult anatomy to embryology the case becomes all the worse for us. Our ear is lodged in the gill-slit of a fish, our jaws are branchial arches, our hyoid bone the rudiment of this system of bones supporting the gills. Our circulation begins as a veritable fish circulation; our earliest skeleton is a notochord; Meckel's cartilage, from which our lower jaw and the bones of our middle ear develop, is a whole genealogical tree of disagreeable ancestors. Our glandula thyreoidea has, according to good authorities, an origin so slimy that it should never be mentioned in polite society. The origin of our kidneys appears decidedly vermian. Time fails me to read merely the name of the witnesses which could be summoned from our own bodies to witness against us.

Even if the testimony of some of these witnesses is not as strong as many think, and we have misunderstood several of them, they are too numerous and their stories hang too well together not to impress an intelligent and impartial jury. But what if it is all true? What if, as some think, our millionth cousin, the tiger or cat, is anatomically a better mammal than I? His teeth and claws and magnificent muscles are of small value compared with man's mental power.

What a comedy that man should work so hard to prove that his chief glory is his opposable thumb, or a few ounces of brain matter! Man's glory is his mind and will, his reason and moral powers, his vision of, and communion with, God. And supposing it be true, as I believe it is true, that the animal has the germ of these also, does that cloud my mind or obscure my vision or weaken my action? It bids me only strive the harder to be worthy of the noble ancestors who have raised me to my higher level and on whose buried shoulders I stand. Whatever may have been our origin, whoever our ancestors, we are men. Then let us play the man. If we will but play our part as well as our old ancestors played theirs, if we will but walk and act according to our light one-half as heroically and well as they groped in the darkness, we need not worry about the future. That will be assured.

Says Professor Huxley: " Man now stands as on a mountain-top far above the level of his humble fellows, and transfigured from his grosser nature by reflecting here and there a ray from the infinite source of truth. And thoughtful man, once escaped from the blinding influences of traditional prejudice, will find in the lowly stock whence man has sprung the best evidence of the splendor of his capacities, and will discern in his long progress through the past a reasonable ground of faith in his attainment of a nobler future."

We have sketched hastily and in rude outline the anatomical structure of the successive stages of man's ancestry ; let us now, in a very brief recapitulation, condense this chronicle into a historical record of progress.

We began with the amœba. This could not have been the beginning. In all its structure it tells us of something earlier and far simpler, but what this earlier ancestor was we do not know. Rather more highly organized relatives of the amœba, the flagellata, have produced a membrane, and swim by means of vibratile, whiplash-like flagella. We must emphasize that these little animals correspond in all essential respects to the cells of our bodies; they are unicellular animals. And the cell once developed remains essentially the same structure, modified only in details, throughout higher animals. And these unicellular animals have the rudiments of all our functions. Their protoplasm and functions seem to differ from those of higher animals only in degree, not in kind. And the more we consider both these facts the more remarkable and suggestive do they become.

Cells with membranes can unite in colonies capable of division of labor and differentiation. And magosphæra is just such a little spheroidal colony. But the cells are still all alike, each one performs all functions equally well. But in volvox division of labor and differentiation of structure have taken place. Certain cells have become purely reproductive, while the rest gather nutriment for these, but are at the same time sensitive and locomotive, excretory and respiratory.

The first function to have cells specially devoted to it is the reproductive; this is a function absolutely necessary for the maintenance of the species. For the nutritive cells die when they have brought the reproductive cells to their full development. These few nutritive cells represent the body of all higher animals in contrast with the reproductive elements. And with the development of a body, death, as a normal process, enters the world. The dominant function is here evidently the reproductive, and the whole body is subservient to this.

In hydra the union and differentiation of cells is carried further. But the cells are still much alike and only slowly lose their own individuality in that of the whole animal. This is shown in the fact that each entodermal cell digests its own particles of food, although the nutriment once digested diffuses to all parts of the body. Also almost any part of the animal containing both ectoderm and entoderm can be cut off and will develop into a new animal.

But beside the reproductive cells and tissues hydra has developed a very simple digestive system, in which the newly caught food at least macerates and begins to be dissolved. This is the second essential function. The animal can, and the plant as a rule does, exist with only the lowest rudiments of anything like nervous or muscular power; but no species can exist without good powers of digestion and reproduction. These essential organs must first develop and the higher must wait. And the inner, digestive, layer of cells persists in our bodies as the lining of the mid-intestine. We compared hydra therefore to a little patch of the lining of our intestine covered with a flake of epidermis; only these layers in hydra possess powers lost to the corresponding cells of our bodies in the process of differentiation. Notice, please, that when cell or organ has once been developed it persists, as a rule, modified, but not lost. Nature's experiments are not in vain; her progress is very slow but sure. But hydra has also the promise of better things, traces of muscular and nervous tissue. There are still no compact muscles, like our own, much less ganglion or brain or nerve-centre of individuality. The tissues are diffuse, but they are the materials out of which the organs of higher animals will crystallize, so to speak. Notice also that these higher muscles and nerves are here entirely subservient to, and exist for, digestion and reproduction.

In the turbellaria the reproductive system has reached a very high grade of development. It is a complex and beautifully constructed organ. The digestive system has also vastly improved; it has its own muscular layers, and often some means of grasping food. But it is slower in reaching its full development than the reproductive system. But all the muscles are no longer attached to the stomach; they are beginning to assert their independence, and, in a rude way, to build a body-wall. But they are in many layers, and run in almost all directions. Some of these layers will disappear, but the most important ones, consisting of longitudinal and transverse fibres, will persist in higher forms. Locomotion by means of these muscles is slowly coming into prominence. They are no longer merely slaves of digestion.

But a muscular fibril contracts only under the stimulus of a nervous impulse. More nerve-cells are necessary to control these more numerous muscular fibrils. The animal now moves with one end foremost, and that end first comes in contact with food, hindrances, or injurious surroundings. Here the sensory cells of feeling and their nerve fibrils multiply. Remember that these neuro-epithelial sensory cells are suited to respond not merely to pressure, but to a variety of the stimuli, chemical, molecular, and of vibration, which excite our organs of smell, taste, and hearing. Such organs and the directive eyes appear mainly at this anterior end. But a ganglion cell sends an impulse to a muscle because it has received one along a sensory nerve from one or more of these sensory cells. Hence the ganglion cells will increase in number. The old cobweb-like plexus condenses into a little knot, the supra-œsophageal ganglion. This ganglion cannot do much, if any, thinking; it is rather a steering organ to control the muscles and guide the animal. It is the servant of the locomotive system. Yet it is the beginning of the brain of higher animals, and probably still persists as an infinitesimal portion of our human brain. And all this is the prophecy of a head soon to be developed. An excretory system has appeared to carry off the waste of the muscles and nerves.

In the schematic worm and annelid the reproductive system is simpler, though perhaps equally effective. It takes the excess of nutriment of the body. The muscular system has taken the form of a sack composed of longitudinal and transverse fibres. The perivisceral cavity, formed perhaps by cutting off and enlarging the lateral pouches of the turbellarian digestive system, serves as a very simple but serviceable circulatory system. But in the annelid and all higher forms a special system of tubes has developed to carry the nutriment, and usually oxygen also, needed to keep up the combustion required to furnish the energy in these active organs. The digestive system has attained its definite form with the appearance of an anal opening and the accompanying division of labor and differentiation into fore-, mid-, and hind-intestine.

The digestive and reproductive systems have thus nearly attained their final form. From the higher worms upward the digestive system will improve greatly. Its lining will fold and flex and vastly increase the digestive and absorptive surfaces. The layer of cells which now secrete the digestive fluids will in part be replaced by massive glands. Far better means of grasping food than the horny teeth of annelids will yet appear. But all these changes are inconsiderable compared with the vast advance made by the muscular and nervous systems. Reproduction and digestion are losing their supremacy in the animal body. Their advance and improvement will require but little further attention.

In the annelid especially, and to some extent in the schematic worm, the supra-œsophageal ganglion is relieved in part of the direct control of the muscular fibrils and has become an organ of perception and the seat of government of lower nervous centres. In all higher forms it innervates directly only the principal sense-organs of the head. And at this stage the light-perceiving directive eye has developed into a form-perceiving, eidoscopic organ. The eye was short of range and its images were perhaps rude and imperfect, but it was a visual eye and had vast possibilities. The animal is taking cognizance of ever more subtle elements in its environment. Perhaps it is not too much to say that the eidoscopic eye first awakened the slumbering animal mind, for its reflex effect upon the supraœsophageal ganglion cannot be over-estimated. The animal will very soon begin to think.

Between the turbellarian and the annelid many aberrant lines diverged. Some of these attained a comparatively high level and then seemed to meet insuperable obstacles, while others came to an end or turned downward very early. Three of these demanded attention, those leading to mollusks, insects, and vertebrates. And it is interesting to notice that the fundamental difference between these three lines was the skeleton, or perhaps we ought to say it was the habit of life which led to the development of such a skeleton.

The mollusk took to a sluggish, creeping mode of life, under an external purely protective skeleton; the insect to a creeping mode of life, with an external but almost purely locomotive skeleton; the vertebrate kept on swimming and developed an internal locomotive skeleton. And it must already have become clear to you that the destiny of these different lines was fixed not so much directly by the skeleton itself as by its reflex effect in moulding the muscular, and ultimately the nervous, system.

The insects formed their skeleton by thickening the horny cuticle of the annelid. They transformed the annelid parapodia into legs and developed wings. They attained life in the air. They devoted the muscles of the body largely to the extremities and gained swift locomotion. They have a fair circulatory and an excellent respiratory system. Best of all, they developed a head and a brain by fusing the three anterior ganglia of the body. The insect could and does think. Such a structure ought to lead to great and high results. But actually their possibilities were very limited. They have not progressed markedly during the last geological period. Their external skeleton was easily attained and brought speedy advantages, which for a time placed them far above all competitors. But it limited their size and length of life and opportunities, and finally their intelligence. They remained largely the slaves of instinct. They followed an attractive and exceedingly promising path, but it led to the bottom of a cliff, not to the summit.

The mollusks, clams, and snails took an easier, downhill road. They formed a shell, and it developed large enough to cover them. It hampered and almost destroyed locomotion and reduced nerve to a minimum. But nerves are nothing but a nuisance anyhow. And why should they move? Food was plenty down in the mud, and if danger threatened, they withdrew into the shell. They stayed down in the mud and let the world go its way. If grievously afflicted by a parasite they produced a pearl—to save themselves from further discomfort. They developed just enough muscle and nervous system to close the shell or drag it a little way; that was all. Digestion and reproduction retained the supremacy. They were fruitful and multiplied, and produced hosts of other clams and snails. The present was enough for them and they had that.

For if the winner in the struggle for existence is the one who gains the most food, the most entire protection against discomfort, danger from enemies or unfavorable surroundings, and the most fruitful and rapid reproduction—and these are all good—then the clam is the highest product of evolution. It never has been surpassed—I venture to say it never can be—except possibly by the tape-worms. I can never help thinking with what contempt these primitive oysters, if they had had brains enough, would have looked down upon the toiling, struggling, discontented, fighting, aspiring primitive vertebrates. How they would have wondered why God allowed such disagreeable, disturbing, unconventional creatures to exist, and thanked him that he had made the world for them, and heaven too, if there be such a place for mollusks. Their road led to the Slough of Contentment.

But even in molluscan history there was a tragic chapter. The squids and cuttle-fishes regained the swimming life, and in their latest forms gave up the protective shell. But its former presence had so modified their structure that any great advance was impossible. It was too late. The sins of the fathers were visited upon the children in the thousandth generation.

The vertebrate developed an internal skeleton. This was necessarily a slow growth, and the type came late to supremacy. The longitudinal muscles are arranged in heavy bands on each side of the back, and the animal swims rapidly. The sense-organs are keen. The brain contains the ganglia of several or many segments and is highly differentiated. It has a special centre of perception, thought, and will; it is an organ of mind. The vertebrate has the physical and mental advantages of large size.

First the definite form and mode of developing a vertebra is attained. Then the vertebral column is perfected. The fins are modified into legs. The lungs increase in size and the heart becomes double.

The animal emerges on land; and, with a better supply of oxygen and less loss of heat, all the functions
are performed with the highest possible efficiency.
First, apparently, amphibia, then reptiles, and finally
mammals of enormous size and strength appeared. It
looked as if the earth were to be an arena where gigantic beasts fought a never-ending battle of brute force.
But these great brutes reproduced slowly, had therefore little power of adaptation, were fitted to special
conditions, and when the conditions changed they disappeared. The bird tried once more the experiment
of developing the locomotive powers to the highest
possible extent. It became a flying machine, and
every organ was moulded to suit this life. Every
ounce of spare weight was thrown aside, the muscles
were wonderfully arranged and of the highest possible
efficiency. The body temperature is higher than that
of mammals. The whole organization is a physiological high-pressure engine. The sense-organs are perhaps the finest and keenest in the whole animal kingdom. The brain is inferior only to that of mammals.
The experiment could not have been tried under more
favorable conditions; it was not a failure, it certainly
was not a success when compared with that of mammals.

The possibilities of every system except one had
been practically exhausted. Only brain development
remained as the last hope of success. Here was an un-
tried line, and the mammals followed it. During the
short tertiary period the brain in many of their genera
seems to have increased tenfold. By the arboreal life
of the highest forms the hand is developed as the instrument of the thinking brain. The battle is begin-

ning to become one of wits, and the crown will soon pass from the strongest to the shrewdest. Mind, not muscle, much less digestion or reproduction, is the goal of the animal kingdom. And we shall see later that the mammalian mode of reproduction and of care of the young led to an almost purely mental and moral advance. For these could have but one logical outcome, family life. And the family is the foundation of society. And family and social life have been the school in which man has been compelled to learn the moral lessons, the application of which has made him what he is.

You must all, I think, have noticed that the different systems of organs succeed one another in a certain definite order; and that each stage from the lowest to the highest is characterized by the predominance of a certain function or group of functions. This sequence of functions is not a deduction but a fact. Place side by side all possible genealogical trees of the animal kingdom, whether founded on comparative anatomy, embryology, palæontology, or all combined. They will all disclose this sequence of functions arranged in the same order. Let me call your attention to the fact that this order is not due to chance, but rests upon a physiological basis. We might almost claim that if the evolution of man from the single cell be granted, no other order of their occurrence is possible.

The protozoa are mostly, though not purely, nutritive and reproductive. These functions are essential to the existence of the species. Naturally in the early protozoan colonies, and in forms like hydra, these functions predominated. But mere digestive tissue is not enough for digestion. Muscles are needed to draw the food to the mouth, to keep the digestive sack in contact with it, and for other purposes. A little higher they are used to enable the animal to go in search of its food. They are still, however, more or less entirely subservient to digestion. But in the highest worms we are beginning to see signs that muscles are predominating in the body; and we feel that, while mutually helpful, the digestive system exists for the muscles, and these latter are becoming the aim of development. From worms upward there is a marked advance in physical activity and strength. The muscles thicken and are arranged in heavier bands. Skeleton and locomotive appendages and jaws follow in insects and vertebrates. The direct battle of animal against animal, and of strength opposed to strength or activity, becomes ever sharper. The strongest and most active are selected and survive.

And yet this is not the whole truth. Some power of perception is possessed by every animal. But until muscles had developed the nervous system could be of but little practical value. Knowledge of even a great emergency is of little use, if I can do nothing about it. But when the muscles appeared, nerves and ganglion cells were necessary to stimulate and control them. And this highest system holds for a long time a position subordinate to that of the lower muscular organ. Its development seems at first sight extraordinarily slow. Only in insects and vertebrates has it become a centre of instinct and thought. Through the sense-organs it is gaining an ever clearer, deeper, and wider knowledge of its environment. First it is affected only by the lower stimuli of touch, taste, and smell. Then with the development of ear and eye it takes cognizance of ever subtler forces and movements. Memory comes into activity very early. The animal begins to learn by experience. The brain is becoming not merely a steering but a thinking organ. More and more nervous material is crowded into it and detailed for its work. Wits and shrewdness are beginning to count for something in the battle. Not only the animal with the strongest muscles, but the one with the best brain survives. And thus at last the brain began to develop with a rapidity as remarkable as its long delay. Thus each higher function is called into activity by the next lower, serves this at first, and only later attains its supremacy.

And yet the advance of the different functions is not altogether successive. Muscle and nerve do not wait for digestion and reproduction to show signs of halting before they begin to advance. They all advance at once. But the progress of reproduction and digestion is most rapid at first, and it appears as if they would outrun the others. But in the ascending series the others follow after, and soon overtake and pass by them. And these lower functions, when outmarched, do not lag behind, but keep in touch with the others, forming the rear-guard and supply-train of the army. And notice that each organ holds the predominance about as long as it shows the power of rapid improvement. The length of its reign is pretty closely proportional to its capacity of development. The digestive system reaches that limit early, the muscular system is capable of indefinitely higher complexity, as we see in our hand. But the muscular system has nearly or quite reached its limit. The body had seen its day of dominance before man arrived on the globe.

But where is the limit to man's mental or moral powers? Every upward step in knowledge, wisdom, and righteousness only opens our eyes to greater heights, before unperceived and still to be attained. These capacities, even to our dim vision, are evidently capable of an indefinite, perhaps infinite, development. What, as yet only partially developed, faculty remains to supersede them? As being capable of an endless development and without a rival, may we not, *must* we not, consider them as ends in themselves? They are evidently what we are here for. Everything points to a spiritual end in animal evolution. The line of development is from the predominantly material to the predominance of the non-material. Not that the material is to be crowded out. It is to reach its highest development in the service of the mind. ' The body must be sustained and perfected, but it is not the end. The goal is mind, the body is of subordinate importance.

But if this is true, we must study carefully the development of mind in the animal. The question presses upon us; if there is a sequence of physical functions in animal development, is there not perhaps also a sequence in the development of the mental faculties? What is the crowning faculty of the human mind and how is its fuller development to be attained? Let us pass therefore to the question of mind in the animal kingdom.

CHAPTER V

WE have sketched hastily the development of the human body. This portion of our history is marked by the successive dominance of higher and higher functions. It is a history treating of successive eras. There is first the period of the dominance of reproduction and digestion, purely vegetative functions, characteristics of the plant just as truly as of the animal. This period extends from the beginning of life up to the time when the annelid was the highest living form yet developed. But in insects and lower vertebrates another system has risen to dominance. This is muscle. The vertebrate no longer devotes all, or the larger part, of its income to digestion and reproduction. If it did, it would degenerate or disappear. The stomach and intestine are improved, but only that they may furnish more abundant nutriment for building and supporting more powerful muscles better arranged. The history of vertebrates is a record of the struggle for supremacy between successive groups of continually greater and better applied muscular power. Here strength and activity seem to be the goal of animal development, and the prize falls to the strongest or most agile. The earth is peopled by huge reptiles, or mammals of enormous strength, and by

8

birds of exceeding swiftness.   This portion of our history covers the era of muscular activity.

But these huge brutes are mostly doomed to extinction, and the bird fails of supremacy in the animal kingdom.   "The race is not to the swift, nor the battle to the strong."   All the time another system has been slowly developing.   The complicated nervous system has required ages for its construction and arrangement. Only in the highest mammals does the brain assert its right to supremacy.   But once established on its throne the brain reigns supreme ; its right is challenged by no other organ.   The possibilities of all the other organs, *as supreme rulers,* have been exhausted. Each one has been thoroughly tested, and its inadequacy proven beyond doubt by actual experiment. These formerly supreme lower organs must serve the higher.   The age of man's existence on the globe is, and must remain, the era of mind.   For the mind alone has an inexhaustible store of possibilities.

The development of all these systems is simultaneous.   From the very beginning all the functions have been represented, all the systems have been gradually advancing.   Hydra has a nervous system just as really as man.   It has no brain, but it has the potentiality and promise of one, and is taking the necessary steps toward its attainment.   But while the development of all is simultaneous, their culmination and supremacy is successive, first stomach and muscle, then brain and mind.   That was not first which is spiritual, but that which is natural ; and afterward that which is spiritual. But now that the mind has once become supreme, man must live and work chiefly for its higher development.   Thus alone is progress possible.

But the word mind calls up before us a long list of powers. And the questions arise, Is one mode and line of mental action just as much the goal of man's development as another? Is man to cultivate the appetite for food and sense gratification just as much as the hunger for righteousness? Or is appetite in the mind like digestion in the body, a function, necessary indeed and once dominant, but no longer fitted for supreme control? Is there in the development of the mental powers or functions just as really a sequence of dominance as in that of the bodily functions? Are there older and lower powers and modes of action, which, though once supreme, must now be rigidly kept down in their proper lower place? Are there lower motives, for which the very laws of evolution forbid us to live, just as truly as they forbid a man's living for stomach or brute strength instead of brain and mind? Are these lower powers merely the foundation on which the higher motives and powers are to rise in their transcendent glory? This is the question which we now must face, and it is of vital importance.

We have come to one of the most important and difficult subjects of zoölogy. Let us distinctly recognize that it is not our task to explain the origin of mind, or even of a single mental faculty. I shall take for granted what many of you will not admit, that the germs of all man's highest mental powers are present undeveloped in the mind, if you will call it so, of the amœba. The limits of this course of lectures have required us to choose between alternatives, either to attempt to prove the truth of the theory of evolution, or taking this for granted, to attempt to find its bearings on our moral and religious beliefs. I have chosen the latter course, and here, as elsewhere, will abide by it. I should not have followed such a course if I did not thoroughly believe that man also, in mind as well as body, is the product of evolution. But this is no reason for your accepting these views. You are asked only to judge impartially of the tendencies of the theory. We take for granted, I repeat, that all man's mental faculties are germinally, potentially, present in protoplasm; we seek the history of their development.

We must remember, further, that the science of animal or comparative psychology is yet in its infancy. Even reliable facts are only slowly being sifted and recorded in sufficient numbers to make deductions at all safe. And even of these facts different writers give very different explanations. As Mr. Romanes has well said, " All our knowledge of mental faculties, other than our own, really consists of an inferential interpretation of bodily activities—this interpretation being founded on our subjective knowledge of our own mental activities. By inference we project, as it were, the human pattern of our own mental chromograph on what is to us the otherwise blank screen of another mind." The value and clearness of our inferences will be proportional to the similarity of the animal to ourselves. Thus we can educate many of our higher mammals by a system of rewards and punishments, and we seem therefore to have good reason to believe that fear and joy, anger and desire, certain powers of perception and inference, are in their minds similar to our own. But fear in a fish is certainly a much dimmer apprehension of danger than in us, even if it deserves the name of apprehension. And the mental state which we call " alarm " in a fly or any lower animal is very difficult to clearly imagine or at all express in terms of our own mind.

Some investigators have made the mistake of projecting into the animal mind all our emotions and complicated trains of thought. Thus Schwammerdam apparently credits the snail with remorse for the commission of excesses. Others go to the other extreme and make animals hardly more than mindless automata. We are warned, therefore, by our very mode of study, to be cautious, not too absolutely sure of our results, nor indignant at others who may take a very different view. And yet by moving cautiously and accepting only what seems fairly clear and evident we may arrive at very valuable and tolerably sure results.

The human mind, and the animal mind apparently, manifests itself in three states or functions. These are intelligence, the realm of knowledge; susceptibility, the realm or state of feelings or emotions; will, the power or state of choice. Let us trace first the development of intelligence or the intellect in the animal. Let us try to discover what kinds of knowledge are successively attained and the mode and sequence of their attainment. Hydra appears to be conscious of its food. It recognizes it partially by touch, perhaps also by feeling the waves caused by its approach. It seems also to recognize food at a little distance by a power comparable to our sense of smell. Stronger impacts cause it to contract. It neither sees nor hears; it probably does little or no thinking. Its knowledge is therefore limited to the recognition of objects either in contact with, or but slightly removed from, itself. And its recognition of the objects is very dim and incomplete, obtained through the sense of touch and smell.

A little higher in the animal world a rude ear has developed, first as a very delicate organ for feeling the waves caused by approaching food or enemies; only later as an organ of hearing. Meanwhile the eye has been developing, to perceive the subtle ether vibrations. The eye of the turbellaria distinguishes only light from darkness, that of the annelid is a true visual organ. Now the brain can begin to perceive the shape of objects at a little distance. Touch and smell, hearing, sight; such is sequence of sense perceptions. The sense-organs respond to continually more delicate and subtle impacts, and cover an ever-widening range of more and more distant objects. Up to this point intelligence has hardly included more than sense-perceptions.

But these sense-perceptions have been all the time spurring the mind to begin a higher work. At first it is conscious merely of objects, and its main effort is to gain a clearer and clearer perception of these.

Now it is led to undertake, so to speak, the work of a sense-organ of a higher grade. It begins to directly see invisible relations just as truly as through the eye it has perceived light. First perhaps it perceives that certain perceptions and experiences, agreeable or disagreeable, occur in a certain sequence. It begins to associate these. It learns thus to recognize the premonitory symptoms of nature's favor or disfavor, and thus gains food or avoids dangers. The bee learns to associate accessible nectar with a certain spot on the flower marked by bright dots or lines, " honey-guides," and the chimpanzee that when a hen cackles there is an egg in the nest. But association is only the first lesson; inference and understanding follow.

The child at kindergarten receives a few blocks. It admires and plays with them. Then it is taught to notice their form. After a time it arranges them in groups and learns the first elements of number. But when it has advanced to higher mathematics, the blocks, or figures on the blackboard, become only symbols or means of illustrating the great theorems and propositions of that science. Thus the animal has begun in the kindergarten way to dimly perceive that there are real, though intangible and invisible, relations between objects. But what is all human science but the clearer vision, and farther search into, and tracing of these same relations? And what is all advance of knowledge but a perception of ever subtler relations? What is even the knowledge of right but the perception of the subtlest and deepest and widest relations of man to his environment? The animal seems to be steadily advancing along the path toward the perception of abstract truth, though man alone really attains it.

And the higher power of association and inference which we call understanding, aided by memory, results in the power of learning by experience, so characteristic of higher vertebrates. The hunted bird or mammal very quickly becomes wary. A new trap catches more than a better old one until the animals have learned to understand it, and young animals are trapped more easily than old. Cases showing the limitations of mammalian intelligence are interesting in this connection. A cat which wished to look out and find the cause of a noise outside, when all the windows were closed by wooden blinds, jumped upon a stand and looked into a mirror. Her inference as to the general use of glass was correct; all its uses had not yet come

within the range of her experience. A monkey used
to stop a hole in the side of a cage with straw. The
keeper, to tease him, used to pull this out. But one
day the monkey tugged at a nail in the side of his cage
until he had pulled it out, and thrust it into the hole.
But when it was pushed back he fell into a rage. His
inference that the nail-head could not be pulled through
was entirely correct; he had failed to foresee that it
could be pushed back. Many such instances have
probably come within the range of your observation, if
you have noticed them. But many of the facts which
Mr. Romanes gives us concerning the intelligence of
monkeys, apes, and baboons would not disgrace the
intelligence of children or men.

Mr. Romanes relates the following account of a little
capuchin monkey from Brazil:

"To-day he obtained possession of a hearth-brush, one of the
kind which has the handle screwed into the brush. He soon
found the way to unscrew the handle, and having done that he
immediately began to try to find out the way to screw it in again.
This he in time accomplished. At first he put the wrong end
of the handle into the hole, but turned it round and round the
right way for screwing. Finding it did not hold he turned the
other end of the handle and carefully stuck it into the hole,
and began again to turn it the right way. It was of course a
difficult feat for him to perform, for he required both his hands
in order to screw it in, and the long bristles of the brush prevented it from remaining steady or with the right side up. He
held the brush with his hind hand, but even so it was very difficult for him to get the first turn of the screw to fit into the
thread; he worked at it, however, with the most unwearying
perseverance until he got the first turn of the screw to catch,
and he then quickly turned it round and round until it was
screwed up to the end. The most remarkable thing was, that
however often he was disappointed in the beginning, he never

was induced to try turning the handle the wrong way; he always screwed it from right to left. As soon as he had accomplished his wish he unscrewed it again, and then screwed it in again the second time rather more easily than the first, and so on many times. When he had become by practice tolerably perfect in screwing and unscrewing, he gave it up and took to some other amusement. One remarkable thing is that he should take so much trouble to do that which is no material benefit to him. The desire to accomplish a chosen task seems a sufficient inducement to lead him to take any amount of trouble. This seems a very human feeling, such as is not shown, I believe, by any other animal. It is not the desire of praise, as he never notices people looking on; it is simply the desire to achieve an object for the sake of achieving an object, and he never rests nor allows his attention to be distracted until it is done. . . .

"As my sister once observed while we were watching him conducting some of his researches, in oblivion to his food and all his other surroundings—'When a monkey behaves like this it is no wonder that man is a scientific animal!'"*

In the highest mammals we find also different degrees of attention and concentration of thought and observation. This difference can easily be noticed in young hunting dogs. A trainer of monkeys said that he could easily select those which could most easily be taught, by noticing in the first lesson whether he could easily gain and hold their attention. This was easy with some, while others were diverted by every passing fly; and the latter, like heedless students, made but slow progress.

It is interesting to notice that one of the perceptions which we class among the highest is apparently developed comparatively early. I refer to the æsthetic perception of the beautiful. Now, the perception of beauty is generally considered as not very far below or

* Romanes: Animal Intelligence, pp. 490, 498.

removed from the perception of truth and right. But some insects and birds apparently possess this perception and the corresponding emotion in no low degree. The colors of flowers seem to exist mainly for the attraction of insects to insure cross-fertilization, and certain insects seem to prefer certain colors. But you may say that these afford merely sense gratification like that which green affords to our eyes or sugar to our tastes.

But does not the grouping of colors in the flower appeal to some æsthetic standard in the mind of the insect? What of the tail of the peacock? Its iridescent rings and eyes evidently appeal to something in the mind of the female. Do form and grouping minister to pure sense gratification? What of the song of the thrush? Does not the orderly and harmonious arrangement of notes and cadences appeal to some standard of order of arrangement, and hence idea of harmony, in the mind of the bird's mate?

Now, I grant you readily that the A B C of this training is mere sense gratification at the sight of bright colors. Most insects and birds have probably not advanced much beyond this first lesson. Savages have generally stopped there or reverted to it. But any appreciation of form and harmonious arrangement of cadence and colors seems to me at least to demand some perception which we must call æsthetic, or dangerously near it. But here you must judge carefully for yourselves lest you be misled. For remember, please, that those schemes of psychology farthest removed from, and least readily reconcilable to, the theory of evolution maintain that perception of beauty is the work of the rational faculty, which also perceives truth and right in much the same way that it perceives and recognizes beauty. If the animal has the æsthetic perception, it has the faculty which, at the next higher stage of development, will perceive, and recognize as such, both truth and right. We are considering no unimportant question ; for on our answer to this depends our answer to questions of far greater importance.

Does it look as if the animal had begun to learn the first rudiments of the great science of rights, of his own rights and those of others ? This is an exceedingly difficult question, though often answered unhesitatingly in the negative. But what of the division of territory by the dogs in oriental cities, a division evidently depending upon something outside of mere brute strength and power to maintain, and their respect of boundaries? The female is allowed, I am told by an eye-witness long resident in Constantinople, to distribute her puppies in unoccupied spots through the city without interference. But when she has once located them, she is not allowed to return and visit them, or pass that way again. So the account by Dr. Washburn of platoons of dogs coming in turn, and peaceably, to feed on a dead donkey in the streets of Constantinople, would seem to be most naturally explained by some dim recognition of rights. Rook communities have not received the attention and investigation which they deserve, but their actions are certainly worthy of attention. Concerning the sense of ownership in dogs and other mammals opinions differ, and yet many facts are most naturally explained on such a supposition.

Just one more question in this connection, for we

are in the borderland or twilightland where it is much safer to ask questions than to attempt to answer them. How do you explain the "instinctive" fear of man on the part of wild and fierce animals? They certainly do not quail before his brute strength, for a blow at such a time breaks the charm and insures an attack. They quail before his eye and look. Is not this the answering of a personality in the animal to the personality in man; a recognition of something deeper than bone and muscle? And may not, as Mr. Darwin has urged, this fear in the presence of a higher personality be the dim foreshadowing of an awe which promises indefinitely better things? Is, after all, the attachment of a dog to his master something far deeper than an appetite for bones or pats, or a fear of kicks?

A host of other and similar questions throng upon us here, to no one of which we can give a definite answer. We need more investigation, more light. We must not rest contented with old prejudices or accept with too great certainty new explanations. The questions are worthy of careful and patient investigation. The study of comparative anatomy has thrown a flood of light on the structure and working of the human body in health and disease. We shall never fully understand the mind of man until we know more of the working of the mind of the animal.

It would seem to be clear that there is a sequence of dominance in the faculties of the intellect. First, the only means of acquiring knowledge is through sense-perception. But memory dawns far down in the animal kingdom. And thus the animal begins to associate past experience with present objects. The bee remembers the gaining of honey in the past, associated with the color of the flower which she now sees, and knows that honey is to be attained again. Thus in time association leads to inference, and understanding has dawned. But the highest faculty of the intellect is the rational intelligence, which perceives beauty, truth, and goodness. This is the last to develop. Traces of its working may be perhaps discovered below man, but only in man does it become dominant. Through it I perceive my rights and duties, and come to the consciousness of my own personality as a moral agent. This tells me of the relation of my own personality to other persons and things. And these are evidently the most important objects of human study. The attainment of this knowledge and the development of this faculty are evidently the goal of human intellectual development. This it is which has insured progress and raised man ever higher above the brutes.

Before we can proceed to the study of the will we must clearly recognize and define certain modes of mental and nervous action, which sooner or later manifest themselves in muscular activity. For, while certain of our bodily activities are clearly voluntary, others take place wholly, or in part independently, of the individual will. Between these different modes of bodily action we must distinguish as clearly as may be possible.

1. Reflex Action. I touch something cold or hot in the dark, suddenly and unexpectedly. I draw back my hand involuntarily and before I have perceived the sensation of cold or heat. You tell me to keep my eyes open while you make a sudden pass at them

with your hand. I try hard to do so, but my eyes shut for all that. I shut them unconsciously and against my own will. I say, "They shut of themselves." Now, this is not true, but the explanation is not difficult. These and similar actions are entirely possible, although the continuity between spinal marrow and brain may have been so interrupted by some accident that sensation in the reflexly active part fails altogether. A bird flaps its wings after its head is cut off, and yet the seat of consciousness and will is certainly in the brain. A patient with a "broken back," and paralyzed in his legs, will draw up his feet if they are tickled, although he is entirely unable to move them by any effort of his will and has no consciousness of the irritation.

The physiological action is in this case clear. The vibration of the nerve caused by the tickling travels from the foot to the appropriate centre in the spinal marrow, and here gives rise to, or is switched off as, a motor impulse travelling back to the muscles of the leg, causing them to contract. In the injured patient the nervous impulse cannot reach the brain, the seat of consciousness, and hence this is not awakened. Normally consciousness does result in a majority of such cases, but only after the beginning or completion of the appropriate action. Yet the movements of our internal organs, intestine and heart, go on continually, and in health we remain entirely unconscious of their action.

But reflex actions may be anything but simple. We walk and talk, and write or play the piano without ever thinking of a single muscle or organ. Yet we had once to learn with much effort to take each step or frame each letter. Thus actions, originally conscious and intended, easily become reflex; often repeated the brain leaves their control to the lower centres. We often say, " I did not intend to do that; I could not help it." We forget that this excuse is our worst condemnation. It is a confession that we have allowed or encouraged a habit to wear a groove from which the wheels of our life cannot escape. The essential characteristic of reflex action is therefore that from beginning to completion it goes on independently of consciousness.

2. Instinct. This is a much-abused word. It is frequently applied to all the mental actions of animals without much thought or care as to its meaning. Let us gain a definition from the study of a typical case lest we use the word as a cloak for ignorance or negligent thoughtlessness. Watch a spider building its wonderful geometrical web. The web is a work of art, and every motion of the spider beautifully adapted to its purpose. But the spider is not therefore necessarily an artist. Let us see of how much the spider is probably conscious, remembering that our best judgment is but an inference. We have good reason to believe that she is conscious of the stimulus to action, hunger. She may be, probably is, conscious of the end to be attained—to catch a fly for her dinner. She seems conscious of what she is doing. In all these respects this differs from reflex action. But she is probably unconscious of the exact fitness of the means to the end. We do not believe that she has adopted the geometrical pattern, because she has discovered or calculated that this will make the closest and largest net for the smallest outlay of labor and

material. Furthermore the young spider builds practically as good a web as the old one. She has inherited the power, not developed or gained it by experience or observation. And all the members of the species have inherited it in much the same degree of perfection.

Concerning the origin of instincts there are several theories. Some instincts would seem to be the result of non-intelligent, perhaps unconscious, habits becoming fixed by heredity and improved by natural selection; others would appear to be modifications of actions originally due to intelligence. Instinct is therefore characterized by consciousness of the stimulus to act, of the means and end, without the knowledge of the exact adaptation of means to end. It is hereditary and characterizes species or large groups.

3. Intelligent Action. You come in cold and sit down before an open fire. You push the brands together to make the fire burn. Applying once more the criterion of consciousness to this action we notice that you are conscious of the stimulus to act, of the steps of the action, and of the end to be attained, exactly as in instinctive action. But finally, and this is the essential characteristic of intelligent action, you are aware to a certain extent of the fitness of the means to the attainment of the end. This piece of knowledge you had to acquire for yourself. Erasmus Darwin defined a fool as a man who had never tried an experiment. Experience and observation, not heredity, are the sources of intelligence. Intelligence is power to think, and a man may be very learned—for do we not have learned pigs?—and yet have very little real intelligence. Hence this is possessed by different individuals in very varying degrees.

We may now briefly compare these three kinds of nervous action.

Reflex action is involuntary and unconscious. The actor may, and usually does, become conscious of the action after it has been commenced or completed, but this is not at all necessary or universal.

Instinctive action is to a certain extent voluntary and conscious. The actor is conscious of the stimulus, the means and mode, and the end or purpose of the action. Of the exact fitness or adaptation of the means to the end the actor is unconscious.

Intelligent action is conscious and voluntary. The actor is conscious of the stimulus to act, of the means and mode, and to a certain extent of the adaptation of the means to the end. This last item of knowledge, lacking in instinctive action, is acquired by experience or observation.

Reflex action may be regarded as a comparatively mechanical, though often very complex, process; the reflex ganglia appear to be hardly more than switch-boards. There is stimulus of the sense-organs, and thus what Mr. Romanes has called "unfelt sensation," unfelt as far as the completion of the action is concerned. But in instinct the sensation no longer remains unfelt; perception is necessary, consciousness plays a part. And this consciousness is a vastly more subtle element, differing as much apparently from the vibration of brain, or nervous, molecules as the Geni from the rubbing of Aladdin's lamp, to borrow an illustration.

But this element of consciousness is one which it is exceedingly difficult to detect in our analysis, and yet upon it our classification and the psychic position of an animal must to a great extent depend. The amœba contracts when pricked, jelly-fishes swim toward the light, the earthworm, " alarmed " by the tread of your foot, withdraws into its hole. Are these and similar actions reflex or instinctive ? A grain of conscious- ness preceding an action which before has been reflex changes it into instinct. Mr. Romanes, probably cor- rectly, regards them as purely reflex. We must, I think, believe that these actions result in conscious- ness even in the lowest forms. The selection and at- tainment of food certainly looks like conscious action. Probably all nerve-cells or nervous material were originally, even in the lowest forms, dimly conscious; then by division of labor some became purely con- ductive, others more highly perceptive. The important thing for us to remember in our present ignorance is not to be dogmatic.

Furthermore, the gain of a grain of consciousness of the adaptation of certain means to special ends changes instinctive action into intelligent, and its loss may reverse the process. Fortunately we have found that in so far as actions, even instinctive, are modified by experience, they are becoming to that extent intel- ligent. This criterion of intelligence seems easily ap- plied. But this profiting by experience must manifest itself within the lifetime of the individual, or in lines outside of circumstances to which its ordinary in- stincts are adapted, or we may give to individual in- telligence the credit due really to natural selection. We must be cautious in our judgments.

These reflex actions are performed independently of consciousness or will. Consciousness may, probably does, attend the selection and grasping of food ; but

most of the actions of the body will go on better without its interference. It is not yet sufficiently developed, or, so to speak, wise enough to be intrusted
with much control of the animal.

Among higher worms cases of instinct seem proven.
Traces of it will almost certainly be yet found much
lower down. Fresh-water mussels migrate into deeper
water at the approach of cold weather. And if the
clam has instincts, there is no reason why the turbellaria should not also possess them. But all higher
powers develop gradually, and their beginnings usually
elude our search. Along the line leading from annelids to insects instinct is becoming dominant. A
supracœsophageal ganglion has developed, and has been
relieved of most of the direct control of the muscles.
Very good sense-organs are also present. From this
time on consciousness becomes clearer, and the brain
is beginning to assert its right to at least know what
is going on in the body, and to have something to say
about it. Still, as long as the actions remain purely
instinctive the brain, while conscious, is governed by
heredity. The animal does as its ancestors always
have. It does not occur to it to ask why it should do
thus or otherwise, or whether other means would be
better fitted to the end in view. It acts exactly like
most of the members of our great political and theological parties. And until the animal has a better
brain this is its best course and is favored by natural
selection.

But the hand of even the best dead ancestors cannot
always be allowed to hold the helm. The brain is
still enlarging, the sense-organs bring in fuller and
more definite reports of a wider environment. Greater

freedom of action by means of a stronger locomotive system is bringing continually new and varied experiences. And if, as in vertebrates, longer life be added, frequent repetition of the experience deepens the impression. Slowly, as if tentatively, the animal begins to modify some of its instincts, at first only in slight details, or to adopt new lines of action not included in its old instincts, but suited to the new emergencies. This is the dawn of intelligence. Its beginnings still remain undiscovered. Mr. Darwin believes that traces of it can be found in earthworms and other annelids. He also tells us that oysters taken from a depth never uncovered by the sea, and transported inland, open their shells, lose the contained water, and die; but that left in reservoirs, where they are occasionally left uncovered for a short time, they learn to keep their shells shut, and live for a much longer time when removed from the water. If oysters can learn by experience, lower worms probably can do the same.

Certain experiments made on sea-anemones, actinæ animals a little more highly organized than hydra, demand repetition under careful observation.* The observer placed on one of the tentacles of a sea-anemone a bit of paper which had been dipped in beef-juice. It was seized and carried to the mouth and here discarded. This tentacle after one or two experiments refused to have anything more to do with it. But other tentacles could be successively cheated. The nerve-cells governing each tentacle appear to have been able to learn by experience, but each group in the diffuse nervous system had to learn separately.

---

* These experiments have been continued with most interesting and valuable results by Dr. G. H. Parker, of Harvard University.

The dawn of this much of intelligence far down in the animal kingdom would not be surprising, for the selection and grasping of food has always involved higher mental power than most of the actions of these lowest animals. Memory goes far down in the animal kingdom. Perhaps, as Professor Haeckel has urged, it is an ultimate mental property of protoplasm. And the memory of past experience would continually tend to modify habit or instinct.

It is unsafe, therefore, to say just where intelligence begins. At a certain point we find dim traces of it; below that we have failed to find them. But that they will not be found, we dare not affirm. In the highest insects instinct predominates, but marks of intelligence are fairly abundant. Ants and wasps modify their habits to suit emergencies which instinct alone could hardly cope with. Bees learn to use grafting wax instead of propolis to stop the chinks in their hives, and soon cease to store up honey in a warm climate.

Our knowledge of vertebrate psychology is not yet sufficient to give a history of the struggle for supremacy between instinct and intelligence, between inherited tendency and the consciousness of the individual. But the outcome is evident; intelligence prevails, instinct wanes. The actions of the young may be purely instinctive; it is better that they should be. But instinct in the adult is more and more modified by intelligence gained by experience. There is perhaps no more characteristic instinct than the habit of nest-building in birds. And yet there are numerous instances where the structure and position of nests have been completely changed to suit new circumstances. And the view that this habit is a pure instinct, unmodified by intelligence, has been disproved by Mr. Wallace. But while size of brain, keenness of sense-organs, and length of life may be rightly emphasized as the most important elements in the development of vertebrate intelligence, the importance of the appendages should never be forgotten. Cats seem to have acquired certain accomplishments — opening doors, ringing door-bells, etc.—never attained by the more intelligent dog, mainly because of the greater mobility and better powers of grasping of the forepaws. The elephant has its trunk and the ape its hand. The power of handling and the increased size of the brain aided each other in a common advance.

The teachableness of mammals is also a sign of high intelligence. The young are often taught by the parent, a dim foreshadowing of the human family relation. And we notice this capacity in domestic animals because of its practical value to man. And here, too, we notice the difference between individuals, which fails in instinct. All spiders of the same species build and hunt alike, although differences caused by the moulding influence of intelligence will probably be here discovered. But among individual dogs and horses we find all degrees of intelligence from absolute stupidity to high intelligence. And many mammals are slandered grievously by man. The pig is not stupid, far from it.

Still only in man does intelligence reign supreme and clearly show its innate powers. But even in man certain realms, like those of the internal organs, are rarely invaded by consciousness, but are normally left to the control of reflex action. These actions go on better without the interference of consciousness.

But other lines of action are relegated as rapidly as

possible to the same control. We learn to walk by a conscious effort to take each step; afterward we take each step automatically, and think only whither we wish to go. We learn by conscious effort to talk and write, to sing, or play the piano. Afterward we frame each letter or note automatically, and think only of the idea and its expression.

So also in our moral and spiritual nature.*

There has been therefore in the successive forms and stages of animal life a clear sequence of dominant nervous actions. The actions of all animals below the annelid are mainly reflex or automatic, unconscious and involuntary. But in insects and lower vertebrates the highest actions at least are instinctive. Conscious-

* Mr. James Freeman Clarke has stated this better than I can. "We may state the law thus: 'Any habitual course of conduct changes voluntary actions into automatic or involuntary (*i.e.*, reflex) actions.' By practice man forms habits, and habitual action is automatic action, requiring no exercise of will except at the beginning of the series of acts. The law of association does the rest. As voluntary acts are transformed into automatic, the will is set free to devote itself to higher efforts and larger attainments. After telling the truth a while by an effort, we tell the truth naturally, necessarily, automatically. After giving to good objects for a while from principle, we give as a matter of course. Honesty becomes automatic; self-control becomes automatic. We rule over our spirit, repress ill-temper, keep down bad feelings, first by an effort, afterwards as a matter of course.

"Possibly these virtues really become incarnate in the bodily organization. Possibly goodness is made flesh and becomes consolidate in the fibres of the brain. Vices, beginning in the soul, seem to become at last bodily diseases; why may not virtues follow the same law? If it were not for some such law of accumulation as this, the work of life would have to be begun forever anew. Formation of character would be impossible. We should be incapable of progress, our whole strength being always employed in battling with our first enemies, learning evermore anew our earliest lessons. But by our present constitution he who has taken one step can take another, and life may become a perpetual advance from good to better. And the highest graces of all—Faith, Hope, and Love—obey the same law." See James Freeman Clarke, Every-Day Religion, p. 122.

ness plays a continually more important part. Still the actions are controlled by hereditary tendency far more than by the will of the individual. But in man instinct has been almost entirely replaced by conscious, voluntary, intelligent action. And yet in man, as rapidly as possible, actions which at first require conscious effort become, through repetition and habit, reflex and automatic. All our conscious effort and the energy of the will, being no longer required for these oft-repeated actions, are set free for higher attainments. The territory which had to be conquered by hard battles has become an integral part of the realm. It now hardly requires even a garrison, but has become a source of supplies for a new advance and march of conquest.

But all this time we have been talking about action and have not given a thought to the will. And we have spoken as if conscious perception and intelligence directly controlled will and action. But this is of course incorrect. Will is practically power of choice. You ask me whether I prefer this or that, and I answer perhaps that I do not care. Until I "care" I shall never choose. The perception must arouse some feeling, if it is to result in choice. I see a diamond in the road and think it is merely a piece of glass. I do not stop. But as I am passing on, I remember that there was a remarkable brilliancy in its flash. It must have been, after all, a gem. My feelings are aroused. How proud I shall feel to wear it. Or how much money I can get for it. Or how glad the owner will be when it is returned to her. I turn back and search eagerly. Perception is necessary, but it is only the first step. The perception must excite some feeling, if choice or exertion of the will is to follow. This is a truism.

Now reflex action takes place independently of consciousness or will. Instinctive action may be voluntary, but it is, after all, not so much the result of individual purpose as of hereditary tendency. Is there then no will in the animal until it has become intelligent? I think there has been a sort of voluntary action all the time. Even the amœba selects or chooses, if I may use the word, its food among the sand grains. And the will is stimulated to act by the appetite. Hunger is the first teacher. And how did appetite develop? Why does the animal hunger for just the food suited to its digestion and needs? We do not know. And the reproductive appetite soon follows. One of these results from the condition of the digestive, the other from that of the reproductive, cells or protoplasm. These appetites are due to some condition in a part of the organism and can be *felt*. They are in a sense not of the mind but of the body. And the response to them on the part of the mind is in some respects almost comparable to reflex action. But the mode of the response is, to a certain extent at least, within the control of consciousness. They train and spur the will as pure reflex action never could. But the will is as yet hardly more than the expression of these appetites. It expresses not so much its own decision as that of the stomach. It is the body's slave and mouthpiece. And once again it is best and safest for the animal that it should be so.

And these appetites are at first comparatively feeble. There is but little muscle or nerve and but little food is required. But these continually strengthen and spur the will harder and more frequently. And the will stirs up the weary and flagging muscles. The

will may be a poor slave and the appetites hard taskmasters. But under their stern discipline it is growing stronger and more completely subjugating the
body. Better slavery to hard taskmasters than rottenness from inertia. The first requirement is power,
activity, and then this power can be directed to ever
higher ends. You cannot steer the vessel until she
has sails or an engine ; with no " way on " she will not
mind the helm, she only drifts. But the condition of
the animal at this stage certainly looks very un-
promising. Can the will emancipate itself from appetite and control it ? Or is it to remain the slave of
the body ?

In time an emotion appears which marks the influence not directly of the body but of the individual
consciousness. This is fear; it is for the body, but
not, like hunger, directly of it. It arises in the mind.
It results from experience and memory. The first
animal which feared took a long step upward. But
when and where was the dawn of fear ? I touch a sea-
anemone and it contracts. Has it felt fear ? I think
not. The action certainly may be purely reflex.
Natural selection, not mind, deserves the credit of that
action. But I am sure that the cat fears the dog, or
the dog the cat, as the case may be. I have little or
no doubt that the bird fears the cat. I am inclined to
believe that the insect fears the bird and the spider
the wasp. But does the highest worm fear? I do
not know. I do not see how there can have been any
fear until there was a nerve-centre highly enough developed to remember past experiences of danger and
fair sense-organs to report the present risk.

Other emotions soon follow. Anger appears early.

The order of appearance of these emotions or motives I shall not attempt to give to you. Indeed this is to us of relatively slight importance. The important point to notice is that a host of these have appeared in mammals and birds, and that each one of these is a new spur to the will. And the will of a horse or dog, to say nothing of a pig, is by no means feeble. And these are slowly emancipating the animal from the tyranny of appetite. But how slow the progress is! Has the emancipation yet become complete in man? I need not answer.

The will has in part, at least, escaped from abject slavery to appetite; it sometimes rises superior to fear. But it is evidently self-centred. The animal may have forgotten the claims of his dead ancestors, he is certainly fully alive to his own interests. Can he even partially rise superior to prudential considerations, as he has to some extent to the claims of appetite? Is it possible to develop the unselfish out of the purely selfish? And if so, how is this to be accomplished? It is not accomplished in the animal; it is but very incompletely accomplished in man. It will be accomplished one day.

In action, at least, the animal is not purely selfish. As Mr. Drummond has shown, reproduction, that old function and first to gain an organ, is not primarily for the benefit of self, but for the species. And not only the storing up of material in the egg, but care for the young after birth, is found in some fish and insects, and increases from fish upward. I readily grant you that this in its beginnings may be purely instinctive, and that not a particle of genuine affection for the young may as yet be present in the mind of the parent.

But beneficial habits may, under the fostering care of selection, develop into instincts. The animal may at first be unconscious of these, and yet they may grow continually stronger. But one day the animal awakens to its actions, and from that time on what had been done blindly and unconsciously is continued consciously, intelligently, and from set purpose. This story is repeated over and over again in the history of the animal kingdom. The care for the young once started as an instinct, affection will follow from the very association of parent with young. Certainly in birds and mammals there seems to be a very genuine love of the parents for their young. This is at first short lived, and the young are and have to be driven away, often by harsh treatment, to shift for themselves. But while it lasts it certainly seems entirely real and genuine. And how strong it is. " A bear robbed of her whelps " is no meaningless expression. And even the weak and timid bird or mammal becomes strong and fierce in defence of her young. In the presence of this emotion appetite and fear are alike forgotten.

But this affection or love once started does not remain limited to parent and offspring. Mammals, especially the higher forms, are social. They frequently go in herds and troops, and appear to have a genuine affection for each other. You all know how in herds of cattle or wild horses the males form a circle around the females and young at the approach of wolves. A troop of orangs were surprised by dogs at a little distance from their shelter. The old male orangs formed a ring and beat off the dogs until the females and young could escape, and then retreated. But as they were now in comparative safety a cry came from one young one, who had been unable to keep up in the scramble over the rocks, and was left on a bowlder surrounded by the dogs. Then one old orang turned back, fought his way through the dogs, tucked the little fellow under one arm, fought his way out with the other, and brought the young one to safety. I call that old orang a hero, but I am prejudiced and may easily be mistaken.

In a cage in a European zoölogical garden there were kept together a little American monkey and a large baboon of which the former was greatly afraid. The keeper, to whom the little monkey was strongly attached, was one day attacked and thrown down by the baboon and in danger of being killed. Then the little monkey ran to his help, and bit and beat his tyrant companion until he allowed the keeper to escape. We are all proud that the little monkey was an American.

Instances of disinterested actions are so common among dogs and horses that farther illustrations are entirely unnecessary. And disinterested action is limited to fewer cases because the environment is rarely suited to its development in the animal world. But do you answer that the affection of the dog is never really disinterested, but a very refined form of selfishness. Possibly. But it were to be greatly desired that selfishness would more frequently take that same refined form among men. But I cannot see how selfishness can ever become so refined as to lead an animal to die of grief over its master's grave.

And if refined selfishness were all, I for one cannot help believing that the dog would long ago have been asleep on a full stomach before the kitchen fire. Has no attempt been made to prove that all human actions are due to selfishness more or less refined? It is very unwise to apply tests and use arguments concerning animals which, if applied with equal strictness to human conduct, would prove human society irrational and purely selfish.

Mammals may be self-centred. But the highest forms have set their faces away from self and toward the non-self; some have at least started on the road which leads to unselfishness.

And man is governed to a certain extent by prudential considerations. If he entirely disregarded these he would not be wise. But the development of the rational faculty has brought before his mind a series of motives higher than these, which are slowly but surely superseding them. Truth, right, and duty are motives of a different order. With regard to these there can be no question of profit or loss. Here the mind cannot stop to ask, Will it pay? Self must be left out of account.

> " When duty whispers low, Thou must,
>   The soul replies, I can."

And thus man rises above appetite, above prudential considerations, and becomes a free and moral agent. And family and social life bring him into new relations, press home upon him new duties and responsibilities, every one of which is a new motive compelling him to rise above self. And thus the unselfish, altruistic emotions have made man what he is, and are in him, ever advancing toward their future supremacy. But some one will say, This is a very pretty theory; it is not history. But the perception of truth and right is certainly a fact, the result of ages of development. And the very highest which the intellect can perceive is bound to become the controlling motive of the will. It always has been so. It must be so, if evolution is not to be purely degeneration. Thus only has man become what he is. And the voice of the people demanding truth and justice, whenever and wherever they see them, is the voice of God promising the future triumph of righteousness. For it is proof positive that man's face is resolutely set toward these, as his ancestors have always marched steadily toward that which was the highest possible attainment.

We find thus that there is a sequence in the motives which control the will. The first and lowest motives are the appetites, and here the will is the mouthpiece of the bodily organs. Then fear and a host of other prudential considerations appear. The lowest of these tend purely to the gratification of the senses or to the avoidance of bodily discomfort. But they originate in the mind, and that is a great gain. But the higher prudential considerations take into account something higher than mere bodily comfort or discomfort. Approbation and disapprobation are motives which weigh heavily with the higher mammals. The lower prudential considerations are purely selfish. The higher ones, which stimulate to action for fellow-animals or men, show at least the dawn of unselfishness. And the altruistic motives, which stimulate to action for the happiness and welfare of others, predominate in, and are characteristic of, man. The human will is slowly rising above the dominance of selfishness. With the dawn of the rational perception of truth, right, and duty, the very highest motives begin to gain

control. And the will becomes more and more powerful as the motives become higher. It is almost a mis-
use of language to speak of the will of a slave of appetite. He is governed by the body, not at all by the
mind.

The man who is governed by prudential considerations, and is always asking, Will it pay? is the incar-
nation of fickleness, instability, and feebleness. The
apparent strength of the selfish will is usually a hollow
sham. But truth, right, and love are motives stronger
than death. And the will, dominated by these, gives
the body to be burned. The man of the future will
have an iron will, because he will keep these highest
motives constantly before his mind.

In the preceding lectures we have traced the sequence of functions and have found that brain and
mind, not digestion and muscle, are the goal of animal
development. In this lecture we have attempted to
trace a corresponding series of functions in the realm
of mind. We have found, I think, that there has been
an orderly and logical development of perceptions,
modes of action, and finally of motives in the animal
mind. Let us now briefly review this history and see
whether it throws any light on the path of man's future
progress.

Most of the sensory cells of the animal minister at
first to reflex action, and there is thus little true perception. The stimuli which have called forth the reflex
action may result afterward in consciousness; but until brain and muscle have reached a higher grade, this
could be of but slight benefit to the animal. Perception and consciousness are exercised mainly in the
recognition and attainment of food. When the animal

begins to show fear, we may feel tolerably certain that it has been conscious of past experience of danger and remembers these experiences. But the sense-organs are all the time improving, whether as servants of conscious perception or of reflex action, and the development of the higher sense-organs, especially of the eyes, has called forth a higher development of the brain. The brain continually develops both through constant exercise and through natural selection. Through the higher and more delicate sense-organs it perceives a continually wider range of more subtile elements in its environment. And the higher the sense-organ the more directly and purely does it minister to consciousness. The eye, when capable of forming an image, is almost never concerned in a purely reflex action.

From the constant recurrence of perceptions and experiences in a constant order the animal begins to associate these, and when he has perceived the one to expect the other. Out of this grows, in time, inference and understanding. The mind is beginning to turn its attention not merely to objects and qualities, but to perceive relations. And thus it has taken the first step toward the perception of abstract truth. And if it has the æsthetic perception and can perceive beauty, we have every reason to believe that the same faculty will one day perceive truth and right. But on the purely animal plane of existence these powers could be of but little service, and we can expect to find them developed only very slightly and under peculiar surroundings. And in this connection it is interesting to notice the great results of man's training and education in the dog. For the wolf and

10

the jackal, the dog's nearest relatives, if not his actual ancestors, are not especially intelligent mammals. Compared with them the dog is a sage and a saint.

The earliest form of action is the reflex. This is independent of both consciousness and will. The only conscious voluntary action of the animal is limited mainly or entirely to the recognition and attainment of food. The motive for the exertion of the will is the appetite, and the will is the slave or mouthpiece of the body. Far higher than this is the stage of instinct. Here the animal is conscious of its actions and new motives begin to appear. But the animal is guided by tendencies inherited from its ancestors. The will has, so to speak, advisory power; it is by no means supreme. But with a wider and deeper knowledge of its environment, with the memory of past experiences, carried by the higher locomotive powers into new surroundings, brought face to face with new emergencies outside of the range of its old instincts, it is compelled to try some experiments of its own. It begins to modify these instincts, and in time altogether does away with many of them. It has risen a little above its old abject slavery to the appetites, it is slowly throwing off the bondage to heredity. New emotions or motives have arisen appealing directly to the individual will. The heir has been long enough under guardians and regents, it assumes the government and can rightly say, "L'état, c'est moi."

But a greater problem confronts it ; can it rise above self? The animal often seems absolutely selfish. Can the unselfish be developed out of the selfish? This seems at first sight impossible. And the first lessons are so easy, the first steps so short, that we do not notice them. Reproduction comes to the aid of mind. The young are born more and more immature. They begin to receive the care of the parent. The love of the parent for the young is at first short lived and feeble. But it is the genuine article, and, like the mustard-seed planted in good soil, must grow. It strengthens and deepens. Soon it begins to widen also. Social life, very rude and imperfect, appears. And the members of this social group support, help, and defend one another. And doing for one another and helping each other, however slightly and imperfectly, strengthens their affection for one another. The animal is still selfish, so is man frequently, but it is in a fair way to become unselfish, and this is all we can reasonably expect of it.

For these are vast revolutions from reflex action to instinct, and from instinct to the reign of the individual will, and from appetite to selfishness on the ground of higher motives, and from immediate gratification to prudential considerations. And the crowning change of all is from selfishness to love. And each one of them takes time. Remember that the Old Testament history is the record of how God taught one little people that there is but one God, Jehovah. Think of the struggles, defeats, and captivities which the Israelites had to undergo before they learned this lesson, and even then only a fraction of the people ever learned it at all. As the prophet foretold, so it came to pass. Though Israel was as the sand by the sea-shore, but a remnant was saved.

But while we seek to do full justice to the animal, let us not underestimate the vast differences between it and man. The true evolutionist takes no low view of man's present actual attainments; in his possibilities he has a larger faith than that of the disbeliever in evolution. In intelligence and thought, in will power and freedom of choice, in one word, in all that makes up character and personality, man is immeasurably superior to the animal. These powers raise him to a new plane of being, give him an indefinitely higher and broader life, and his appearance marks a new era. He alone is a moral, responsible being, to a certain extent the former of his own destiny and recorder of his doom, if he fails. This gives to all his actions a peculiar stamp of a dignity only his. What he is and is to be we must attempt to trace in another lecture. But to one or two characteristic results of his progress we must call attention here.

The principal subject of man's study is not so much the things which surround him as his relation to them and theirs to each other. His environment has become really one, not so much one of tangible and visible objects as of invisible relations. And these will demand endless investigation. The more he studies them the more wonderful do they become. The vein broadens and grows indefinitely richer the deeper he searches into it. We find thus the purpose of the intellect; it is to study environment.

And now a little about motives. The animal begins with appetite, and some animals and men never get any farther. And yet how easily this appetite for food is satiated! We all remember our experiences as children around the Thanksgiving or Christmas table. What a disappointment it was to us to find how soon our appetite had forsaken us, and that we had lost the power of enjoying the delicacies which we had most anticipated. And over-indulgence often brought sad results and was followed by a period of penitential fasting. And the appetites for sense gratification must always lead to this result. They not only crave things which " perish with the using ; " temporarily at least, often permanently, the appetite itself perishes with the gratification.

But what of the appetite, if you will pardon the expression, for truth and right? All attainment only strengthens it; and, instead of enslaving, it makes men ever more free. And yet what a power there is in the appetite for truth and righteousness? In obedience to it man gives his body to be burned, or pours out his life-blood drop by drop for its attainment, and rejoices in the sacrifice. There are victims to appetite : there are only martyrs to truth. This soul hunger for truth and right, growing more intense as the soul is filled with the object of desire, is the only one capable of indefinite development and dominance of the will. This must be and is the mental goal of animal development, if man has a future corresponding in length at all to his past. Otherwise the history of life becomes a "story told by an idiot." For its satisfaction is the only one which never causes satiety, and of which over-indulgence is impossible. All others lead only to a slough of despond, or the deeper and more treacherous slough of contentment, beyond which rise no delectable mountains or golden city.

And now in closing let me call your attention to one thought of practical vital importance.

According to the theory which we have agreed to adopt, higher species have arisen through a process of natural selection, those species surviving which are best conformed to their environment. And this applies to man as well as to lower animals. All knowledge is in man, therefore, primarily, a means by which he may conform to environment, survive, and progress. But conformity includes more than mere knowledge of environment. A man might have all knowledge, and yet refuse to conform; and then his knowledge could not save him from destruction. For conformity alone gives survival. Conformity in man requires an effort of the will. It is intelligent, but it is also voluntary action. And knowledge is a necessary means of conformity because through it we see how we may conform, and because it furnishes the motives which stimulate the will to the necessary effort.

Now, that faculty of the intellect which is dominant in man, and which has raised him immeasurably above the animal, and made him man, is the rational intelligence. If there is any such thing as a law of history or as continuity in evolution, man's future progress must depend upon his clearer vision and recognition of the perceptions of this faculty. Through it man perceives beauty, truth, and goodness, and attains knowledge of himself as a person and moral agent, and recognizes his rights and duties. Of all this the animal is and remains unconscious; indeed he is not yet a moral being and person in any proper sense of the word.

Inasmuch as the rational perception is the dominant faculty in man, it must perceive the lines along which he is to conform. Truth, right, and duty must be his watchwords. These are to be the rules and motives of all his actions. He cannot live for the body, but for something higher, the mind. This was proven before man appeared on the globe. He is to be a mental, intelligent being. But he is not to be governed by appetite or mere prudential considerations. These are animal, not human motives. These are not to be disregarded any more than digestion can be safely disregarded by man. But they are not to be his chief motives. He must subordinate these to the higher motives furnished by right and duty. Man is not merely a mental but a moral being. If he sinks below this plane of life he is not following the path marked out for him in all his past development. In order to progress, the higher vertebrate had to subordinate everything to mental development. In order to become man it had to develop the rational intelligence. In order to become higher man, present man must subordinate everything to moral development. This is the great law of animal and human development clearly revealed in the sequence of physical and mental functions.

Must man be a religious being also? This question we must try to answer in a future lecture.

CHAPTER VI

NATURAL SELECTION AND ENVIRONMENT

I HAVE attempted to show that animal development has not been an aimless drifting. Functions developed and organs arose and were perfected in a certain order. First the purely vegetative organs appeared, and the animal lived for digestion and reproduction; then came muscle and it brought with it nerve. But these were not enough; the brain had all the time been gradually improving, and now it becomes the dominant function to which all others are subordinated. The experiment was fairly tried. Mere digestion and reproduction are carried to about the highest perfection which can be expected of them in worms and mollusks. The bird tried what could be done with digestion ministering to locomotion guided by the very keenest sense-organs and controlled by no mean brain. Even this experiment was not a success. But one organ remained, the brain, and on its mental possibilities depend the future of the animal kingdom. Vegetative organs and muscle have been tried and found wanting.[1]

We have followed hastily the development of mind. The mind began its career as the servant of digestion, recognizing and aiding to attain food. Action is at first mainly reflex. But conscious perception plays an ever more important part. The animal is at first guided by natural selection through the survival of

[1] See chart, p. 310.

the most suitable reflex actions, then by inherited tendencies, finally by its own conscious intelligence and will. The first motives are the appetites, but these are succeeded by ever higher motives as the perceptions become clearer and more subtile relations in environment are taken into account. Governed first purely by appetites, the will is ever more influenced by prudential considerations, and finally shows well-developed " natural affections." It has set its face toward unselfishness.

Digestion and muscle, as well as mind, have persisted in man. He is not, cannot be, disembodied spirit. And in his mental life reflex action and instinct, appetite and prudence, are still of great importance. But the higher and supreme development of these powers could never have resulted in man. They might alone have produced a superior animal, never man. His mammalian structure found its logical and natural goal in family and social life. And even the lowest goal of family life is incompatible with pure selfishness, and as family life advanced to an ever higher grade it became the school of unselfishness and love. And social life had a similar effect.

Moreover, man as a social being early began to learn that he could claim something from his fellows, and that he owed something to them. If he refused to help others, they would refuse to help him. This was his first, very rude lesson in rights and duties. Love, duty, and right have ever since been the watchwords of his development and progress. We have not yet considered, and must for the present disregard, the value and efficiency of religion in aiding his advance. At present we emphasize only the historical fact that man has not become what he is by a higher development of the body, nor by giving free rein to appetite, nor yet by making the dictates of selfish prudence supreme. And if there is any such thing as continuity in history, such modes and aims of life, if now followed, would surely only brutalize him and plunge him headlong in degeneration. He must live for right, truth, love, and duty. In just so far as he makes any other aim in life supreme, or allows it to even rival these, he is sinking into brutality. This is the clear, unmistakable verdict of history, and we shall do well to heed it.

But granting all that can be claimed for this sequence, have not the lower forms whose anatomy we have sketched—worm, fish, and bird—halted at various points along this line of march? Yet they have evidently survived. And if they have found safe resting-places, cannot higher forms turn back and join them? In other words, is not degeneration easier than advance and just as safe? What is the result if an animal tries to return to a lower plane of life or refuses to take the next upward step? Generally extermination. The very classification of worms in a number of small isolated groups, which must once have been connected by a host of intermediate forms, is indisputable proof of most terrible extermination. They did not go forward, and the survivors are but an infinitesimal fraction of those which perished. Let us take an illustration where palæontology can help us. The earth was at one time covered with marsupial mammals. Some advanced into placental forms. The great mass remained behind. And outside of Australia the opossums are the only survivors of them all. And this is only one example where a thousand could be given. Place is not long reserved for mere cumberers of the ground. There are so few exceptions to this statement that we might almost call it a law of biology.

Let us see how it fares with an animal which retreats to a lower plane of life. A worm, rather than seek its own food, becomes a parasite. It degenerates, but still is easily recognized as a worm. A crustacean tries the same experiment, though living outside of its host instead of in it. It sinks to a place even lower, if possible, than that of the parasitic worm. A locomotive form becomes sessile. It loses most of its muscles and the larger part of its nervous system ; and even the digestive system, which it has made the goal of its existence, is inferior to that of its locomotive ancestors and relatives. But to the vertebrate these lowest depths of stagnation and degeneration are, as a rule, impossible. From true fish upward parasitism and sessile life are practically impossible. Here stagnation and degeneration mean, as a rule, extinction. Of all the relatives of vertebrates back to worms only the very aberrant lines of amphioxus and of the tunicata remain. Of the rest not a single survivor has yet been discovered. And yet what hosts of species must have peopled the sea. The primitive round-mouthed fishes have practically disappeared. The ganoids survive in a few species out of thousands. The amphibia of the carboniferous and the next period and the reptiles of the mesozoic have disappeared ; only a few feeble degenerate remnants persist. And this was necessarily so. Each advancing form crowded hardest on those which occupied the same place and sought the same food, that is, the members of the same species. And the first to suffer from its competition were its own brethren. Death, rarely commuted into life imprisonment, is the verdict pronounced on all forms which will not advance. And does not the same law of advance or extinction apply to man? What is the record of successive civilizations but its verification?

Notice once more that as we ascend in the scale of development natural selection selects more unsparingly and the path to life narrows. It is a very easy matter for the lowest forms to get food. Indeed the plant sits still and its food comes to it. And the battle of brute force can be fought in a multitude of ways— by mere strength, by activity, by offensive or defensive armor, or even by running into the mud and skulking. It is harder to gain knowledge, and yet many roads lead to an education. Colleges are by no means the only seats of education. And many totally uneducated men have college diplomas. And life is, after all, the great university, and here the sluggard fails and the plucky man with the poor " fit " often carries off the honors.

> "But where shall wisdom be found?
> And where is the place of understanding?
> The gold and the crystal cannot equal it:
> And the exchange of it shall not be for jewels of fine gold.
> No mention shall be made of corals or of pearls:
> For the price of wisdom is above rubies."

And when it comes to righteousness there is only one right, and everything else is wrong. " Wide is the gate and broad is the way that leadeth to destruction, and many there be that go in thereat: Because strait is the gate and narrow is the way which leadeth unto

life, and few there be that find it." Therefore " strive
to enter in at the strait gate." And remember that
"strive " means wrestle like one of the athletes in the
old Olympic games.

" I saw also that the Interpreter took Christian again by the
hand and led him into a pleasant place, where was built a
stately palace beautiful to behold ; at the sight of which Christian was greatly delighted. He saw also, upon the top thereof,
certain persons walking, who were clothed all in gold. Then
said Christian, May we go in thither ?

"Then the Interpreter took him and led him up toward the
door of the palace ; and, behold, at the door stood a great company of men, as desirous to go in, but durst not. There also
sat a man at a little distance from the door at a table-side, to
take the name of him that should enter therein; he saw also
that in the door-way stood many men in armour, to keep it,
being resolved to do to the men that would enter what hurt and
mischief they could. Now was Christian somewhat in amaze.
At last, when every man started back for fear of the armed
men, Christian saw a man of a very stout countenance come up
to the man that sat there to write, saying, Set down my name,
Sir; the which when he had done, he saw the man draw his
sword, and put an helmet upon his head, and rush toward the
door upon the armed men, who laid upon him with deadly
force ; but the man, not at all discouraged, fell to cutting and
hacking most fiercely. So after he had received and given
many wounds to those that attempted to keep him out, he cut
his way through them all, and pressed forward into the palace,
at which there was a pleasant voice heard from those that were
within, even of those that walked upon the top of the palace
saying :

" 'Come in, come in ;
Eternal glory thou shalt win.'

" So he went in, and was clothed in such garments as they.
" Then Christian smiled, and said, I think verily I know the
meaning of this."—Bunyan's, Pilgrim's Progress, p. 44.

If you wish to climb the Matterhorn many paths
lead up the lower slopes, and a stumble here may
cost you only a sprain. And I suppose that several
paths lead to the base of the cone. But thence to the
summit there is but one path, and a misstep means
death. Pardon these quotations and illustrations.
They are my only means of at all adequately present-
ing to you a scientific man's conception of the meaning
of the struggle for life. The laws of evolution are
written in blood and bear the death penalty. For

"Life is not as idle ore,
But iron dug from central gloom,
And heated hot with burning fears,
And dipt in baths of hissing tears,
And battered with the shocks of doom
To shape and use."

There would seem therefore to be going on a process
of natural selection. Natural selection seems to select
more unsparingly and the struggle for life—or even
existence—to grow fiercer as we advance from lower
forms to higher in the animal kingdom.

But the theory which we have agreed to accept
teaches us that these survivors are those which or who
have conformed to their environment and that they
have survived because of their conformity. And what
do we mean by environment? And does not man
modify his environment? Certainly he changes by
irrigation a desert into a garden. He carries water
against its tendency to the hill-top. But he has learned
to do this only by studying the laws which govern the
motions of fluids and rigorously obeying them. He
must carry his water in strong pipes and take it from

some higher point, or must use heat or some means to furnish the force to drive it to the higher point. He cannot change a single iota of the law, and gains control of the elements only by obedience to their laws. Electricity is man's best servant as long as he respects its laws, but it kills him who disobeys them. But does not man make his own surroundings in social life? He merely enters upon a new mode of life; and if this new mode be in conformity with the eternal forces and laws of environment man prospers in this new mode of life and conforms still more closely.

There is, indeed, but one environment, but the lower animal comes in contact with, and is affected by, but a small portion of its elements. Form and color were in the world before the animal had developed an eye, but up to this time these could have but little effect on animal life. Light vibrations were present in ether long before the animal by responding to them made them any part of its own true environment. There is vastly more in environment than man has yet discovered, and he will discover these elements only by obedience to their laws.

Environment includes ultimately all the forces and elements which go to make up our world or universe. It is an exceedingly general term. I might say that under the environment of certain wheels, springs, and spindles, which we call a Jacquard loom, silk threads become a ribbon worthy of a queen. Is Nature and environment only a huge divine loom to weave man and something higher yet? One great difference is evident. Under normal conditions the silk must become a ribbon. But protoplasm can fail to conform and become waste. Environment is a very hard word to define, and our views concerning it may differ.

One thing, however, seems to me clear and evident. If each successive stage in the ascending series is selected or survives on account of its conformity to environment there must be some element or power, something or somewhat in environment specially corresponding in some way to, or suited to drawing out, the characteristic of this ascending stage on account of which it survives. The forces and elements of environment make and work against those at each stage who wander from the right path, and for those who follow it. And thus natural selection arises as the total result of the combined working of all these forces. They all unite in one resultant working along a certain line, and natural selection is the effect of this resultant. In the stage represented by hydra the forces of environment combine in a resultant which works for digestion and reproduction and the best development of their organs. But as the animal changes he comes into a new relation or occupies a new position in respect to these forces. New elements in the old environment are beginning to press upon him. And the resultant changes accordingly. He may be compared to a steamer at sea which raises a sail. The wind has been blowing for hours, but the sail gives it a new hold on the ship. Steam and wind now combine in a new resultant of forces. From worms upward environment manifests itself through natural selection as a power working for muscular force and brute strength or activity.

But soon natural selection ceases to select on the ground of brute force. After a time environment proves to be a power making for shrewdness. And when the mammal has appeared the resultant of the forces of environment impels more and more toward unselfishness, and when man has appeared environment proves to be a " power, not ourselves, that makes for righteousness." But what shall we say of an environment which unmasks itself at last as a power making for intelligence, unselfishness, and righteousness? Someone may answer it is a host of chemical and physical forces bringing about very high ends. That is very true, but is it the whole truth? The thinking man must ask, How did it come about, and why is it that all these forces work together for such high moral and intelligent ends?

We face, therefore, the question, Can an environment which proves finally and ultimately to be a power not ourselves making for righteousness and unselfishness be purely material and mechanical? Or must there be in or behind it something spiritual? Shall we best call environment, in its highest manifestation, "it" or "him?"

The old argument of Socrates, as on the last day of his life he sits discoursing with his friends, still holds good. He is discussing the same old question, whether there is anything more than force, material, mechanism in the world. He says that one might assign as "the cause why I am sitting here that my body is composed of bones and muscles; that the bones are solid and separate, and that the muscles can be contracted and extended, and are all inclosed in the flesh and skin; and that the bones, being jointed, can be drawn by the muscles, and so I can move my legs as you see; and that this is the reason why I am sitting here. But by

11

the dog, these bones and muscles would long ago have carried me to Megara or Bœtia, moved by my opinion of what was best, if I had not thought it more right and honorable to submit to the sentence pronounced by the state than to run away from it. To call such things causes is absurd. For there is a great difference between the cause and that without which the cause would not produce its effect."

If there is no intelligence or love of truth in the cause, how can there be anything higher in the effect? And if Socrates had been only bone and muscle, he ought to have run away.

Our problem stands somewhat as follows : We have given protoplasm, a strange substance of marvellous capacities, which we call functions, and possessing a power of developing into beings of ever higher grades of organization. Environment proves to be a combination of forces working for the higher development of functions in a certain orderly sequence. And every lower function in the ascending line demands the development of the next higher. Digestion demands muscle, and muscle nerve, and nerve brain. We shall soon see that mammalian structure had to culminate in the family, and the family demands unselfishness and obedience. Environment therefore proves from the beginning to have been unceasingly working for the highest end ; never, even temporarily, merely for the lower. For we have seen that environment works most unsparingly against those who, having taken certain of the steps in the ascending path, fail to continue therein.

But in order to attain this highest end for which it has always been working, an immense number of subsidiary ends have had to be attained. These are not merely digestion and brain, but a host of others : *e.g.*, in vertebrates, vertebræ of the right substance, position, form, arrangement, and union. And in the ascending line, for whose highest forms it has continually worked, the difficulties of attaining each subsidiary end have been successively solved, and through this host of subsidiary ends the animal kingdom has advanced straight to its goal of intelligence and righteousness. Now the whole process is a grand argument for design. But I would not emphasize the process so much as the end attained. This especially, when attained by conformity to that environment, demands more than mere mindless atoms in or behind that environment. Can we call the ultimate power which makes for righteousness "it?" Can we call it less than "Him, in whom we live and move and have our being?"

The history of life is a grand drama. "Paradise Lost" and Shakespeare's plays are but fragments of it. But without intelligence they could never have been composed ; without a choice of means and ends they could never have been placed upon the stage. Does the plot of this grander drama of evolution demand no intelligence in its ultimate cause and producer? Is the succession of steps, each succeeding the other in such order as to lead to truth and right and continual progress toward a spiritual goal, is this plot possible without a great composer who has seen the end from the beginning? Could it ever have been executed upon the stage of the world, and perhaps of the universe, without an executing will?

Now I freely grant you that this is no mathematical

demonstration. Natural science does not deal in
demonstrations, it rests upon the doctrine of probabilities ; just as we have to order our whole lives according to this doctrine. Its solution of a problem is
never the only conceivable answer, but the one which
best fits and explains all the facts and meets the fewest objections. The arguments for the existence of a
personal God are far stronger than those in favor of
any theory of evolution. But we very rightly test the
former arguments, indefinitely more rigidly and severe-
ly, just because our very life hangs on them. On the
other hand, we should not reject them as useless, because they are not of an entirely different kind from
those on which all the actions and beliefs of our common daily life are based. There is a scepticism which
is merely a credulity of negations. This also we
should avoid.

We have considered a few of the reasons for thinking that, with the material, there must be something
spiritual in environment, that if the woof is material
the warp is God. Here we need not delay long.
Blank atheism seems to be at present unpopular and
generally regarded as unscientific. The so - called
philosophic materialism of the present day seems to
be in general far nearer to pantheism than to the old
form of materialism which recognized only atoms and
mechanism. Atheism as a power to deform the lives
of men has, for the present, lost its hold, and even
agnosticism is respectful. The materialism against
which we have to struggle is not that of the school,
but of the shop, of society, of life. There are comparatively few now who avow a system of philosophy
making mindless atoms their first cause.

But there is a far grosser, more deadly materialism of the heart and will. It sits unrebuked in the front pews of our churches and controls alike church and parish, caucus and legislature. It calls on us all to fall down and worship, promising the world if we obey, the cross if we refuse. And we bow to it; and that is all it asks, for a nod on our part makes us its slaves. It is the idolatry of money, position, shrewdness, learning—in one word, of success. It takes all the strength out of our morality, loyalty and obedience to God out of our religion, and makes cowards and liars of us, who should be heroes. It makes our religion a byword with honest unbelievers. And if they are honest scientific minds, waiting for evidence of the practical value of our religion, why should they believe, when we live so successfully down to the religion which we would scorn to openly profess? Our fathers may have been narrow or straight-laced; they were not cross-eyed from trying to keep one eye on God and the other on the main chance. What is the use of whispering, "Lord, Lord," Sundays, if we shout, "Oh, Baal, hear us," all the rest of the week. Let us at least be honest, and "if Baal be god, follow him," and avow it. And worst, and most hideous, of all, we are not so much hypocrites as self-deceived. Let us not forget the old Greek doctrine of Ate, goddess of judicial blindness, sent down only upon those who were living the unpardonable sin of indifference.

But supposing that there is in environment something more and other than material, can we possibly know anything about it?

I am in a boat near the mouth of a river. The boat is tossed by the waves, driven by currents of wind, and now and then temporarily turned by eddies. I seem to look out upon a chaos of apparently conflicting forces. But all the time the wind and tide are sweeping me homeward. Now the wind, which sometimes indeed does shift, and the great tidal wave are steadily bearing me in a certain direction, though wave and eddy and gust may often make this appear doubtful to me. So, underneath all waves and eddies of environment, there is a great tidal wave, bearing man steadily onward; and I gain a certain amount of valid knowledge of environment from the direction in which it is bearing me.

Let us change the illustration. Man survives as all his ancestors have survived before him, through conformity to environment. Environment has therefore during ages past been continually making impressions upon him. And he can draw valid inferences concerning the one 'power, which must underlie the apparent host of forces of environment, from the impressions which these have left upon the structure of his mind and character. By studying himself he gains valid knowledge of what is deepest in environment. For man is the most completely and closely conformed thereto of all living beings.

But man *is* a religious being. This is a fact which demands explanation just as much as bone and muscle. Now no evolutionist would believe that the eye could ever have developed without the stimulus of light acting upon the cells of the skin. Place the animal in darkness and the eye becomes rudimentary and disappears. Could a visual organ for seeing moral and religious truth have ever originated in the mind of man had there been no corresponding pulsation and thrill of a corresponding reality in environment? Is not the one development just as improbable or inconceivable as the other?

And this is the reason that, when man awakened to himself and his own powers, he knew that there was and must be a God. "Pass over the earth," says Plutarch; "you may discover cities without walls, without literature, without monarchs, without palaces and wealth; where the theatre and the school are not known; but no man ever saw a city without temples and gods, where prayers and oaths and oracles and sacrifices were not used for obtaining pardon or averting evil." Given man and environment as they are, and a belief in God is a necessary result. But you may ask, if we are to worship a personal God, why might not a conscious and religious hydra, with equal right, worship an infinite stomach, and the annelid a god of mere brute force?

There stands in Florence a magnificent statue by Michel Angelo. A human figure is only partially hewn out of the stone. He never finished it. If you could have seen the master hewing the chips. with hasty, impatient blows from the shapeless block, you would have been tempted to say that he was but a stonecutter, and but a hasty workman at that. Even now we do not know exactly what form and expression he would have given to the still unfinished head. But no one can examine it and hesitate to pronounce it a grand work of a master-mind. In any manifestly incomplete work you must judge the purpose and character and powers of the workman or artist by its highest possibilities, just so far as you have any reason to believe that these possibilities will be realized. You must look at the rudely outlined heroic human figure in the block of stone, not at the rough unfinished pedestal, if you would know Michel Angelo. So in the hydra and the annelid you must look at the possibilities of the nervous system before you or he think that digestion and muscle are all.

Once more the highest powers dawn far down in the animal kingdom. There are traces of mind in the amœba, and of unselfishness in the lower mammals. If there were a goal of human development higher and other than unselfishness, wisdom, and love, we should have seen traces of it before this. But have we found the faintest sign of any such? Moreover, remember that a function continues to develop about as long as it shows the capacity for development. And during that period environment is a power making for its higher development. But is there any limit to the possible development of the three mental activities mentioned above? I can see none. Then must we not expect that environment will always make for these? And will environment ever manifest itself to man as the seat or instrument of a power possessing higher faculties other than these? Man must worship a personal God of wisdom, unselfishness, and love, or cease to worship. The latter alternative he never yet has been able to take, and society survive under its domination. So I at least am compelled to read the finding of biological history.

But let us grant for the sake of argument that man contains still undeveloped germs of faculties capable of perceiving and attaining something as much higher than wisdom and love as these are higher than brute force. You will answer, this is not only inconceivable, it is impossible. Still let us grant the possibility. We notice, first of all, that it is against the whole course of evolution that these faculties should be other than mental, and what we class under powers pertaining to our personality. For ages past evidently, and no less really from the very beginning, evolution has worked for the body only as a perfect vehicle of mind, and for this as leading to will and character. And human development has led, and ever more tends, as Mr. Drummond has shown, to the arrest, though not the degeneration, of the body. It is to remain at the highest possible stage of efficiency as the servant of mind. These higher powers will thus·be mental and personal powers. And how has any and every advance to higher capabilities been attained in the animal kingdom? Merely by the most active possible exercise of the next lower power. This is proven by the sequence of physical and mental functions. We shall attain, therefore, any higher mental capacities only by the continual practice of wisdom and love. That is our only path to something higher, if higher there shall ever be. But if we find that the God of our environment is a God of something higher than love and righteousness, will these cease to be characteristics of his nature and essence? Not at all.

I have learned, perhaps, to know my father as a plain citizen. If I later find that he is a king and statesman, with powers and mental capacities of which I have never dreamed, do I therefore from that time cease to think of him as wise and kind and good? Not in the least. I only trust his love and wisdom as guide of my little life all the more. And shall not the same be true of God though he be king of all worlds and ages ? It becomes unwise and wrong to worship God as the God of might only when we have found that he is a God also of something higher and nobler, of love ; and after we have perceived this fully and worship him as love, we rest in the arms of his infinite power.

But now that the work has gone thus far, we can see that all development must take place along personal, spiritual lines; and are compelled to believe in a spiritual cause who knew the end from the beginning. And man's farther progress depends upon his conformity to this spiritual environment. And what is conformity to the personal element in our environment but likeness to him ? This is my only possible mode of conformity to a person—to become like him in word, action, thought, and purpose, and finally in all my being. Very far from a close resemblance we still are. But we are more like him than primitive man was; and our descendants will resemble him far more closely than we. And thus man, conscious of his environment, and that means capable of knowing something about God, knows at least what God requires of him, namely, righteousness, love, and likeness to himself; or, as the old heathen seer expressed it, "to do justly, love mercy, and walk humbly before God." Man is and must be a religious being. And he conforms consciously. Thus to be more like God he must know more about him, and to know more about him he must become more like him. The two go hand in hand, and by mutual reaction strengthen each other. I will not enter into the most important question of all, whether we can ever really know a person unless we have some love for him. The facts of evolution seem to me to admit of but one interpretation, that of Augustine : " Thou hast formed me for thee, O Lord, and my restless spirit finds no rest but in thee." Granted, therefore, a personal God in and behind environment, however dimly perceived, and conformity to environment means god-likeness ; for conformity to a person can mean nothing less than likeness to him.

Some of you must, all of you should, have read Professor Huxley's " Address on Education." In it he says, " It is a very plain and elementary truth that the life, the fortune, and the happiness of every one of us, and, more or less, of those who are connected with us, do depend upon our knowing something of the rules of a game infinitely more difficult and complicated than chess. It is a game which has been played for un-known ages, every man and woman of us being one of the two players in a game of his or her own. The chess-board is the world, the pieces are the phenomena of the universe, the rules of the game are what we call the laws of Nature. The player on the other side is hidden from us. We know that his play is always fair, just, and patient. But also we know, to our cost, that he never overlooks a mistake, or makes the smallest allowance for ignorance. To the man who plays well the highest stakes are paid with that sort of over-flowing generosity with which the strong shows delight in strength. And one who plays ill is checkmated—without haste, but without remorse.

" My metaphor," he continues, " will remind some of you of the famous picture in which Retzsch has depicted Satan playing at chess with man for his soul.

Substitute for the mocking fiend in that picture a calm, strong angel, who is playing for love, as we say, and would rather lose than win—and I should accept it as an image of human life." [1]

This is a marvellous illustration, and in general as true as it is beautiful and grand. But that "calm, strong angel who is playing for love, as we say, and would rather lose than win," is certainly a very strange antagonist. Is it, after all, possible that our clear-eyed scientific man has altogether misunderstood the game? Is not the "calm, strong angel" more probably our partner? Certainly very many things point that way. And who are our antagonists? Look within yourself and you will always find at least a pair ready to take a hand against you, to say nothing of the possibilities of environment. "Rex regis rebellis." Our partner is trying by every method, except perhaps by "talking across the board," to teach us the laws and methods of this great game. And calls and signals are always allowable. The game is not finished in one hand; he gives us a second and third, and repeats the signals, and never misleads. Only when we carelessly or obstinately refuse to learn, and wilfully lose the game beyond all hope, does he leave us to meet our losses as best we may.

Let us carry the illustration a step farther. Who knows that the game was, or could be, at first taught without talking across the board? I can find nothing in science to compel such a belief, many things render it improbable. Grant a personality in environment to which personality in man is to conform and gain likeness. Environment can act on the digestive and mus-

---

[1] Huxley: Lay Sermons and Addresses, p. 31.

cular systems through mere material. But how can personality in environment act on personality in man except by personal contact or by symbols easy of comprehension according to its own laws? Some method of attaining acquaintance at least we should certainly expect. .

But some of you may ask, How can any theory of evolution guarantee that anything of the present shall survive in the future? It is continually changing and destroying former types. The old order of everything changes and passes away, giving place to the new. But is this the whole truth? Evolution is a radical process, but we must never forget that it is also, and at the same time, exceedingly conservative. The cell was the first invention of the animal kingdom, and all higher animals are and must be cellular in structure. Our tissues were formed ages on ages ago; they have all persisted. Most of our organs are as old as worms. All these are very old, older than the mountains, and yet I cannot doubt that they must last as long as man exists. Indeed, while Nature is wonderfully inventive of new structures, her conservatism in holding on to old ones is still more remarkable. In the ascending line of development she tries an experiment once exceedingly thorough, and then the question is solved for all time. For she always takes time enough to try the experiment exhaustively. It took ages to find how to build a spinal column or brain, but when the experiment was finished she had reason to be, and was, satisfied. And if this is true of bodily organs we should expect that the same law would hold good when the animal development gradually passes over into the spiritual. And what is human history but the record of moral and religious experiments, and their success or failure according as the experimenters conformed to the laws of the spiritual forces with which they had to do?

We need not fear that our old fundamental beliefs will be lost. Their very age shows that they have been thoroughly tested in the great experiment of human history and found sure. Modified they may be; they will be used for higher purposes and the building of better characters than ours. They will not be lost or discarded. We too often think of nature as building like man, with huge scaffoldings, which must later be torn down and destroyed. But in the forest the only scaffolding is the heart of oak.

We have seen that the sequence of functions in animal development has culminated in man's rational, moral nature. He alone has the clear perception of the reality of right, truth, and duty. The pursuit of these has made him what he is. His advance, if there is any continuity in history, depends upon his making these the ruling motives and aims of his life. He must continually grow in righteousness and unselfishness, if he is not to degenerate and give place to some other product of evolution. Moreover, as these moral faculties are capable of indefinite, if not infinite, development, they must dominate his life through a future of indefinite duration. For the length of the period of dominance of a function has always been proportional to the capacity of that function for future development. These can never, so far as we can see, be superseded, for no rival to them can be discovered. We have found in them the culmination of the sequence of functions.

We have attempted to show in this lecture that reversal of this grand sequence has always led to degeneration, or, in higher forms, far more frequently, to extinction. As we ascend, natural selection works more, rather than less, unsparingly. And as advance depends upon conformity to environment, and as the highest forms must be regarded as therefore most completely conformed, we gain our most adequate knowledge of environment when we study it as working especially for these. For these have been from the very beginning its far-off, chief aim and goal. Viewed from this standpoint, environment proves to be a host of interacting forces uniting in a resultant " power, not ourselves, that makes for righteousness," and unselfishness.

Inasmuch as man's rational moral nature, his personality, is the result of the last and longest step toward and in conformity to environment, these powers correspond to that which is at the same time highest, and deepest, and most fundamental in that environment. This power which makes for righteousness is therefore to be regarded as personal and spiritual rather than material. It is God immanent in nature. And it is mainly to this personal and spiritual element in his environment that man is in the future to more completely conform. Conformity to this element in man's environment does not so much result in life as it *is* life ; failure to conform is death. And the pressure of environment upon man, compelling him to choose between life through conformity and non-conformity with death, can be most naturally and adequately explained as the expression of his will. We know what he requires of us.

Our knowledge of him is very incomplete, but may be valid as far as it extends. And it would seem to be valid, for it has been tested by ages of experiment. The results of this grand experiment have been summed up in man's fundamental religious beliefs. And farther knowledge will be gained by more complete obedience to the requirements already known. The evidence, that these fundamental religious beliefs will persist, is of the same character as that upon which rests our belief in the persistence of cells and tissues. The one is rooted in the structure of our minds; the other, in the structure of our bodies. But, after all, only will can act upon will, and personality upon personality. It remains for us to examine how man was compelled by his very structure to develop a new element in his environment, conformed indeed to the laws of his old environment, but better fitted to draw out the moral and spiritual side of his nature. And in connection with this study we may hope to gain some new light on the laws of conformity.

CHAPTER VII

WE are too prone to think that soil and climate, hillside or plain, mountain and shore, temperature and rainfall, constitute the sole or the most important elements in human environment. Every one of these elements is doubtless important. Frost, drought, or barrenness of soil may make a region a desert, or dwarf the development of its inhabitants. Mountaineer, and the dweller on the plain, and the fisherman on the shore of the ocean develop different traits through the influence of their surroundings. In too warm a climate the human race loses its mental and moral vigor and degenerates. This is undeniable.

But, though one soil and climate and set of physical surroundings may be more conducive than another to the development of heroism, truthfulness, unselfishness, and righteousness, no one is essential to their production or sure to give rise to them. Moral and religious character is a feature of man's personality, and our personality is moulded mainly by the men and women with whom we associate. A man is not only "known by the company which he keeps;" he is usually fashioned by and conforms to it. As President Seelye has well said, "The only motive which can move a will is either a will itself, or something into which a will enters. It is not a thought, but only a

12

sentiment, a deed, or a person, by which we become
truly inspired. It is not the intellect, but the heart
and will, through which and by which we are controlled. It is not the precepts of life, but life itself, by which alone we are begotten and born unto
life.

"Now, there are two ways in which living power,
personal power, the power of a will, may enter a soul
and give it life ; the one is when God's will works upon
us, and the other when our wills work upon one another. God's will may directly penetrate ours, enabling
us to will and to do of his good pleasure ; and our own
wills, thus inspired, may be the torch to kindle other
wills with the same inspiration. It is in only one of
these two ways that a human soul can be truly inspired ; and, without a true inspiration, no amount of
instruction, whether in duty, or life, or anything else,
will change a single moral propensity." *

Even though a Lincoln may rise above his hereditary position or his surroundings, they are the school
in which he is trained ; the gymnasium in which his
mental and moral fibre is strengthened. Family and
social life form thus the element of man's environment
by which he is mostly moulded, and to which he most
naturally and completely conforms. Let us therefore
briefly trace the origin of this new element of man's
environment, and then notice the effect upon him of
conformity to its laws, and see whither these would
lead him.

We have already seen that intra-uterine development of the young was being carried ever farther by
mammals, and we found one explanation of this in the

* Seelye : Christian Missions, p. 154.

fact that each mammalian egg represented a large amount of nutriment, and that the mammal had very little material to spare for reproduction. Very possibly, too, the newly hatched mammals were exposed to even more numerous and greater dangers than the young of birds. Even among lower mammals the young is feeble at birth. But the human infant is absolutely helpless. And the centre of its helplessness is its brain. Its eyes and ears are comparatively perfect, but its perceptions are very dim. Its muscles are all present, but it must very slowly and gradually learn to use them. Its language is but a cry, its few actions reflex. The new-born kitten may be just as helpless, but in a few weeks it will run and play and hunt, and after a few months can care for itself. Not so the child. It must be cared for during months and years before it can be given independence. Its brain is so marvellously complex that it is finished as a thinking and willing and muscle-controlling mechanism only long after birth. This means a period of infancy during which the young clings helplessly to the mother, who is its natural protector. And during this period the mother and young have to be cared for and protected by the male. And the period of infancy and the protection of the female and young are just as truly, though in far less degree, characteristic of the highest apes as of man.

I can give you only this very condensed and incomplete abstract of Mr. John Fiske's argument; you must read it for yourself in his "Destiny of Man." And as he has there shown, this can have but one result, and that is the family life of man. And we may yet very possibly have to acknowledge that family life of a very low grade is just as truly characteristic of the higher apes as of lower man. And thus the family life of man is the physiological result of, and rooted in, mammalian structure.

And the benefits of family life are too great and numerous to even enumerate. First of all the family is the school of unselfishness. All the love of the parent is drawn out for the helpless and dependent child, and grows as the parent works and thinks for it. And the child returns a fraction of his parents' love. Within the close bond of the family the struggle for place and opportunity is replaced by mutual helpfulness; and this doing and burden-bearing with and for each other is a constant exercise in the practice of love. And without this mutual love and helpfulness the family cannot exist.

And slowly man begins to apply the lessons learned in the family to other relations with partners, neighbors, and friends. Slowly he discovers that an entirely selfish life defeats its own ends. A voice within him tells him continually that love is better than selfishness and ministering better than being ministered unto. It dawns upon him that it is against the nature of things that other people should be so selfish and grasping; a few begin to apply the moral to themselves, and a few of these to act accordingly.

And what a change the few steps which man has taken in this direction have wrought in his life. Says Professor Huxley: " In place of ruthless self-assertion it demands self-restraint, in place of thrusting aside or treading down all competitors, it requires that the individual shall not merely respect, but shall help his fellows ; its influence is directed not so much to the survival of the fittest as to the fitting of as many as possible to survive. It repudiates the gladiatoral theory of existence."

It is a vast change from the "gladiatorial theory" to that of "mutual helpfulness." Call it a revolution, if you will. Revolutions are not unheard of in the history of the animal kingdom any more than in human history. We have seen, first, digestion and reproduction on the throne of animal organization, then muscle, and finally brain. Each of these changes is in one sense a revolution.

A little before the summer solstice the earth is whizzing away from the sun; a few weeks later it is whizzing with equal rapidity in almost the opposite direction. In the very nature of things it could not be otherwise. But so silently and gradually does it come about that we never feel the reversal of the engine; indeed the engine has not been reversed at all. Very similar is the change of the struggle of brute against brute to that of man for man. Indeed human development seems now to be almost at such a solstice where the power that makes for love is almost exhausted in opposing the tendency toward selfishness. We shall not always stay at the solstice; soon we shall make more rapid progress. And unselfishness like the family relation is firmly rooted in mammalian structure.

And man owes almost everything to family life. First the child gains the advantage of the parent's experience. He is educated by the parent. In a few formative and receptive years he gains from the parent the results of centuries of human experience. The process is thus cumulative, the investment bears compound interest. And yet this is peculiar to man only in degree. Have you never watched a cat train her kittens? And the education of the child in the savage family is very incomplete.

The family is the first and fundamental of all higher social and political unities. And without the persistence of the family the larger social unit would become an inert mass. All the individual ambition, all desire for family advancement, must be retained as still a motive for energetic advance. And all the training which social life can give reaches the individual most effectively, or solely, through the family. Society without the family would be like an army without company or regimental organization. Thus the very existence, not only of training in love and mutual helpfulness, but even of society itself as a mere organization, depends upon the existence and improvement of family life. And as so much depended upon and resulted from it, it could not but be fostered and improved by natural selection. The tribe or race with the best family life has apparently survived. But all social animals have some means of communicating very simple thoughts or perceptions. The simplest illustrations of this are the calls and warning cries of mammals and birds. It is not impossible that the higher mammals have something worthy of the name of language. But man alone, with his better brain and better anatomical structure of throat and mouth, and the closer interdependence with his fellows, has attained to articulate speech. And this again has become the bond to a still closer union.

Now our only question is, How does social life enable and aid man to conform to environment? We are interested not so much in his happiness as in his progress.

It helps and improves the body by giving him a better and more constant supply of more suitable food, and better protection from inclemency of the weather, and in many other ways. Baths and gymnasia are built, and medical science prolongs life. Yet make the items as many as you can, and what a long list of disadvantages to man physically you must set over against these. Many of these evils will doubtless disappear as society becomes better organized, but some will always remain to plague us. We pamper or abuse our stomachs, and dyspepsia results. We live in hot-houses, and a host of diseases are fostered by them. Indeed it would be hard to count up the diseases for which social life is directly or indirectly responsible. Social life becomes more and more complicated, and our nervous systems cannot bear the strain. Medical science saves alive thousands who would otherwise die, and these grow up to bear children as weak as themselves. We are looking now at the physical side alone ; and from this standpoint the survival of the invalid is a sore evil. Now society will and must become healthier ; we shall not always abuse our bodies as sinfully as we now do. Still, viewed from the standpoint of the body alone, the best, as it seems to me, which we can claim, is that social life does no more harm than good.

What has social life done for man intellectually? Much. It gives him schools and colleges. But are our systems of education an unmixed good? How many of our schools and colleges are places where men are stuffed with facts until they have no time nor inclination to think? They may turn out learned men ; do they produce thinkers? And how about the spread of knowledge? Is it not a spread of information ? And most of what goes forth from the press is not worthy of even that name, or is information which a man had better be without. We are proud of being a nation of readers. And reading is good, if a man thinks about what he reads ; otherwise it is like undigested food in the stomach, an injury and a curse. A dyspeptic gourmand is helped by "cutting down his rations." In our mental disease we need the same course of treatment. Let us read fewer books and papers and think more about what we do read.

Society may foster original thinking; it is none the less opposed to it.

> " Yon Cassius has a lean and hungry look,
> He thinks too much ; such men are dangerous."

This is the motto of all great parties in Church and State. Still social life has undoubtedly fostered thought. We think vastly more and better than primitive man ; still we have much to learn. Society puts the experience of centuries at the service of every individual. Poor and unsatisfactory as are our modes of education, they are a great blessing intellectually and will become more helpful. And, after all, the friction of mind against mind in social life—provided social intercourse is this, and not the commingling of two vacua—is a continual education of inestimable advantage. And all these advantages would without language have been absolutely impossible. Intellectually our debt to society is inestimable.

And how does social life aid man morally ? I cannot help believing that primitive society was the first school of the human conscience. It was a rude school, but it taught man some grand lessons.

The primitive clan would seem to have existed as a rude army for the defence of its members and for offensive operations against enemies. Individual responsibility on the part of its members was slight for offences against individuals of other clans, or against the gods. For any such offence of one of its members the whole clan was held, or held itself, largely responsible. If one man sinned, the clan suffered. It could not therefore afford to pardon wilful disobedience to regulations made by it or its leaders. Its very existence depended on this strict discipline. And much the same stern discipline has to be maintained in our modern armies or they become utterly worthless.

Furthermore, man, as a social being, is very ready to accept the estimate of his actions placed upon them by his fellows. It is not easy to resist public opinion now. The tie of class or professional feeling is a tremendous power for good and evil. It must have been almost irresistible in that primitive army, which summarily outlawed or killed the obstinately disobedient. But all obedience was lauded and rewarded. It had to be so. And if the tribe was worthy to survive, because its regulations were better than those of its rivals, or perhaps as nearly just and right as were well possible, it was altogether best and right it should be so. The voice of the people was, in a very rude, stammering way, the voice of God. And those who survived became more and more obedient, and found themselves, when disobedient, feeling debased, and mean, and unworthy, as their fellows considered them. And all this feeling tended to develop a conscience in the individual answering to the estimates and regulations of the community.

And remember that the primitive religion is a tribal religion. The gods felt toward a man just as his neighbors did. A public opinion of this sort is irresistible, and a man's conscience and estimate of himself and his actions must confrom to it. But you may say a man may grant that this opinion is in a sense irresistible, and find himself very miserable and unhappy under its condemnation. But he would not feel remorse; this is a very different feeling. Possibly it may be. I am not so sure. But what I am interested in maintaining is that the condemnation of one's fellowmen puts more vividly before one's eyes, and emphasizes, the condemnation of one's own self. It may often be a necessary step in self-conviction. And what is most important, even in our own case, the condemnation of our fellows often brings with it self-condemnation.

Try the experiment, as you will some day, of following a course of action which you feel fairly confident is right, but which all your neighbors think is foolish and wrong. See if you do not feel twinges within you which you must examine very closely to distinguish from twinges of conscience. If you do not, I see but one explanation—you are conscious that God is with you, and content with this majority. But in the case of primitive man God was always on the side of one's tribe.

Now this does not explain the origin of man's conception of right; it presupposes such a conception in some dim form. I do not now know why right is right or beauty beautiful. I only know they are so. Where or when either of these perceptions dawned I do not know. But, given some such dim perception, I believe that primitive human society gave it its iron grip on every fibre of man's nature.

Before the animal could safely be allowed to govern itself intelligently it had to serve a long apprenticeship to reflex action and instinct. And man's moral nature had to undergo a similar apprenticeship to tribal regulation and tribal conscience. Only slowly was instinct modified and replaced by intelligent action. And how this old tribal conscience persists. Often for good, although there it were better replaced by an individual conscience working for right. But how slowly you and I learn that there is a higher responsibility than to party or class. How often my vote and action are controlled, not by my own conscience, but by the opinion of my fellows, or the feeling that, if my party suffers defeat, God's work will suffer at the hands of my opponents. And what is all this but the survival in a very degenerate form of the old tribal conscience of primitive man? And he knew, and could know, nothing better: I can and do.

But society slowly works for unselfishness. The love learned in the family manifests itself in ever-widening circles ; it must do so if it is the genuine article. It works for neighbors and friends, then for the poor and helpless of the community. Then it spreads to other communities and nations. For genuine love recognizes no bounds of time or place. Slowly we learn that we are our brother's keepers, and that the brotherhood cannot stop short of the human race. Goodness and kindness radiate from one, perhaps unknown, member of the community to his fellows, and thence all over the world. And the world is the better for his one action.

Primitive society was thus the best possible school of conscience; and the family and it are the great school of unselfishness. But society is even more and better than this. It is the medium through which thought, power, and moral and religious life can spring from man to man. This is its last and culminating advantage: it is that for which society really exists.

For, in the close bonds of family and social life, a new possibility of development has arisen based upon articulate speech. We might almost call it a new form of heredity, independent of all blood-relationship. Progress in anatomical structure in the animal kingdom was slow, because any improvement could be transmitted only to the direct descendants of its original possessor. But in all matters pertaining to or based upon mind, a new invention, or idea, or system becomes the property of him who can best appreciate it. The torch is always handed on to the swiftest runner. Thus Socrates is the true father of Plato, and Plato of Aristotle. Whoever can best understand and appreciate and enter into the spirit of Socrates and Plato becomes heir to their thoughts and interprets them to us. And the thought of one man enriches all races and times.

But a great teacher like Socrates is not merely an intellectual power. "Probe a little deeper, surgeon," said the French soldier, "and you'll find the emperor." Napoleon may have impressed himself on the soldier's intellect; he had enthroned himself in his heart. "Slave," said the old Roman, Marius, to the barbarian who had been sent into the dungeon to despatch him, "slave, wouldst thou kill Caius Marius?" And the barbarian, though backed by all the power of Rome,

is said to have fled in dismay.   Why did he run away ?
I do not know.   I only know that I should have done
the same.   One more instance.   Some thirty years
ago the northern army was fleeing, a disorganized mob,
toward Winchester.   Early had fallen upon them suddenly in the gray of the morning, and, while one corps
still held its ground, the rest of the army was melting
away in panic.   Then a little red-faced trooper came
tearing down the line shouting, " Face the other way
boys; face the other way."   And those panic-stricken
men turned and rolled an irresistible avalanche of
heroes upon the Confederate lines.   What made them
turn about?   It was something which I can neither
define nor analyze—the personal power of Sheridan.
It is the secret of every great leader of men.   Now
Sheridan had imparted more than information to these
men.   Is it too much to say that he put himself into
them ?   From such men power streams out like electricity from a huge dynamo.

Now society furnishes the medium through which
such a man can act.   You have all met such men,
though probably not more than one or two of them.
But one such man is a host.   They may be men of few
words.   But their very presence and look calls out all
that is good in you ; and while you are with them evil
loses its power.   Says the gay and licentious Alci-
biades, in Plato's " Banquet " concerning Socrates :

"When I heard Pericles or any other great orator,
I was entertained and delighted, and I felt that he
had spoken well.   But no mortal speech has ever excited in my mind such emotions as are excited by this
magician.   Whenever I hear him, I am, as it were,
charmed and fettered.   My heart leaps like an in-

spired Corybant. My inmost soul is stung by his words as by the bite of a serpent. It is indignant at its own rude and ignoble character. I often weep tears of regret and think how vain and inglorious is the life I lead. Nor am I the only one that weeps like a child and despairs of himself. Many others are affected in the same way."

These men are the real kings. Their power for good, and sometimes for evil, is inestimable. And the great advantage of social life, as a means of conforming to environment, is the medium which it furnishes to conduct the power of such men. Man's last effort toward conformity to environment, the struggle for existence in its last most real form, is the life and death grapple between good and evil. For here good and evil, righteousness and sin, come face to face in spiritual form; "we wrestle not with flesh and blood." Life is more than a game of chess or whist; it is a great battle; every man must, and does, take sides; he must fight or die. And the real kings of society are, as a rule, on the side of truth, and aid its triumph. For one essential condition of such leadership is the power to inspire confidence in the love of the king for his willing subject. A suspicion of selfish aims in the leader breaks this bond. The hero must be self-forgetful. This is one reason for man's hero-worship, and the magnetic, dominant power of the hero. But evil is essentially selfish and can gain and hold this kingship only as long as it can deceive. And these kings "live forever." Dynasties and empires disappear, but Socrates and Plato, Luther and Huss, Cromwell and Lincoln, rule an ever-widening kingdom of ever more loyal subjects.

And society will have leaders; men may set up whatever form of government they will, they are always searching for a king. And this is no sign of weakness or credulity. Man's desire for leadership is only another proof of the vast future which he knows is before him, and into which he longs to be guided. The wiser a man is, the more he desires to be taught; the nobler he becomes, the more whole-souled is the homage which he pays to the noblest. Is it a sign of weakness or ignorance in students, of adult age and ripe manhood, to flock to some great university to hear the wisdom and catch the inspiration of some great master? When Jackson fell Lee exclaimed, "I have lost my right arm." Was Jackson any the less for being the right arm to deal, as only he could, the crushing blows planned by the great strategist?

But is not man to be independent and free? Certainly. But he gains freedom from the petty tyranny of robber-baron or boss, and from the very pettiest tyranny of all, the service of self, only as he finds and enlists under the king. Serve self and it will plunge you in, and drag you through, the ditch, till your own clothes abhor you. You are free to choose your teacher and guide and example. But choose you will and must. I am not propounding theories; I am telling you facts. Whether for better or worse man always does and will choose because he must. Look about you, look into yourselves. Have you no hero whom you admire and strive to resemble? no teacher to whom you listen? You must and do have your example and teacher. Is he teaching you to conform to environment, or leading you to be ground in pieces by its forces all arrayed against you?

The Carpenter of Nazareth stood before Pilate. "And Pilate said unto him, Art thou a king then? Jesus answered, Thou sayest that I am a king. To this end was I born, and for this cause came I into the world, that I should bear witness unto the truth. Every one that is of the truth heareth my voice." And Pilate would not wait for the answer to his question, What is truth? and the Jews chose Barabbas. Would you and I have acted differently? The answer of our Lord to Pilate contains the essence of Christianity. "You a king," says Pilate in astonishment; "where is your power to enforce your authority?" And our Lord's answer seems to me to mean substantially this: Roman legions shall suffer defeat, rout, and extermination; and Roman power shall cease to terrify. All its might must decay. But "everyone that is of the truth" shall attach himself to me with a love which will brave rack and stake. All your power cannot give a grain of new life. I can and will infuse my own divine life, my own divine *self*, into men. And this new life is invincible, immortal, all-conquering. I have infused myself into a few fishermen, and they will infuse *me* into a host of other men. Thus I will transfigure into my own character every man in the world, who is of the truth, and therefore will hear my voice. All the power of Rome cannot prevent it, and whatever opposes it must go down before it.

Christianity is the contagion of a divine life. Society is the medium through which it could and was to work. Greece had prepared the language necessary for its spread. Roman power had built its highways and levelled all obstructions.

" A little leaven leaveneth the whole lump." " Not by might, nor by power, but by my Spirit, saith the Lord of Hosts."

But, you will object, the grandest kings have had, as a rule, the fewest loyal subjects. The prophets and seers are stoned. Elijah stands alone on Carmel and opposed to him are more than a thousand prophets of Baal, with court and king at their head. Heroism does not pay, and heroes are few. Right is always in a hopeless minority. Let us look into this matter carefully, for the objection, even if overstated, certainly contains a large amount of truth.

Let us go back to two forms having much the same grade of organization: both worms. One of them sets out to become a vertebrate, building an internal skeleton. The other forms an external skeleton and becomes a crab. To form its skeleton the crab had only to thicken the cuticle already present in the annelid. It had to modify the already existing parapodia and their muscles, changing them to legs. The external skeleton gave from the start a double advantage —protection and better locomotion. Every grain of thickening aided the animal in the struggle for existence in both these ways. The very fact that the skeleton was external may have rendered it more liable to variation, because it was thus exposed to continual stimuli. And the best were rapidly sifted out by Natural Selection. The change and development went on with comparative rapidity. In the mollusk the change was apparently still more easy and the development still more rapid.

But the development of an internal skeleton was more difficult and slower. It was of no use for the

13

protection of the animal, and only gradually did it
become of much service in locomotion. Being deep-
seated it very possibly changed all the more slowly.
Furthermore, a cartilaginous rod, like the notochord,
even fully developed, hardly enabled the animal to
fight directly with the mail-clad crab. The internal
skeleton had to become far more highly developed before its great advantages, and freedom from disadvantages, became apparent. The mollusk and crab were
working a mine rich in surface deposits although
soon exhausted. The vertebrate lead was poor at the
surface, and only later showed its inexhaustible richness. It looked as if the vertebrate were making a
very poor speculation.

Whether this explanation be true or not, a glance
at a chart, showing the geological succession of occurrence of the different kingdoms, proves that in the
oldest palæozoic periods there were well-developed
cuttlefish and crabs before there were any vertebrates
worthy of the name. If any were present, their skeleton was purely cartilaginous and not preserved.

I think we may go farther, although in this latter
consideration we may very possibly be mistaken.
We have already seen that the progress made by any
animal may be measured more or less accurately by
the length of time during which its ancestors maintained a swimming life. The ancestors of the cœlenterates settled to the bottom first. Then successively
those of flatworms, mollusks, annelids, and crabs. All
this time the ancestors of vertebrates were swimming
in the water above. Food was probably more abundant, certainly more easily and economically obtained
by a creeping life, on the bottom. But thither the

vertebrate could not go. There his mail-clad competitors were too strong for him. Those which settled and tried to compete in this sort of life perished. We may have to except the ascidia, but they paid for their success by the loss of nearly all their vertebrate characteristics. The future progress of vertebrates depended upon their continual activity in the swimming life. And they were forced by their environment to maintain this. Otherwise they might, probably would, never have attained their present height of organization. Certainly at this time you would have found it hard to believe that the victory was to fall to these weaker and smaller vertebrates.

Let us come down to a later period. Reptiles, mammals, and birds are struggling for supremacy. Of the power and diversity of form of these old reptiles we have generally no adequate conception. The forms now living are but feeble remnants. There were huge sea-serpents, and forms like our present crocodiles, but far more powerful. Others apparently resembled in form and habit the herbivorous and carnivorous mammals of to-day. Others strode or leaped on two legs. And still others flew like bats or birds. They were terrible forms, with coats of mail and powerful jaws and teeth. And they were active and swift. When we look at them we see that the vertebrate, though slow in gaining the lead, is sure to hold it. The internal skeleton gave fewer advantages at the start; its greatest superiority had lain in future possibilities.

But which vertebrate is heir to the future? It would have been a hard choice between reptile and bird. I feel sure that I, for one, should not have se-

lected the mammal, a small, feeble being, hiding in
holes and ledges, and continually hard put to it to
escape becoming a mouthful for some huge reptile.
And yet the persecution, the impossibility of contend-
ing by brute strength, may have forced the mammal
into the line of brain-building and placental development. The early development of mammals appears
to have been slow. Palæontology proves that they
were long surpassed by reptiles and birds. But the
little mammal had the future. The battle was to go
against the strong.

Once again. The arboreal life of higher mammals
would seem to be most easily explained by the view
that they were driven to it by stronger carnivorous
mammals having possession of the ground. Brain
was good, for it planned escape from enemies. But
it did not give its possessor immediate victory over
muscle, tooth, and claw in the tiger. That was to
come far later with the invention of traps and guns.
Brain gave its possessor a sure hold of the future, and
just enough of the present to enable it to survive by
a hard struggle. And the same appears to have been
true of primitive man.

Thus all man's ancestors have had to lead a life of
continual struggle against overwhelming odds and of
seeming defeat. It was a life of hardship, if not of
positive suffering. The organ which was to give them
future supremacy, whether it was backbone, placenta,
or brain, could in its earlier stages aid them only to a
hardly won survival. The present apparently, and
really as far as freedom from discomfort and danger is
concerned, always belongs to forms hopelessly doomed
to degeneration or stagnation. Crabs, not primitive

vertebrates, were masters of the good things of the sea; and, in later times, reptiles, not mammals, of those of the land. Any progressive form has to choose between the present and the future. It cannot grasp both. I am not propounding to you any metaphysical theories, but plain, dry, hard facts of palæontology; explain them as you will.

And here we must add our last word about conformity to environment; and it is a most important consideration. Conformity to environment is not such an adaptation as will confer upon an animal the greatest immunity from discomfort or danger, or will enable it to gain the greatest amount of food and place, and produce the largest number of offspring. Indeed, if you will add one element to those mentioned above, namely, that all these shall be attained with the least amount of effort, they insure degeneration beyond a doubt. This is the conformity of the bivalve mollusk. The clam has abundance of food, enormous powers of reproduction, almost perfect protection against enemies, and lives a life of almost absolute freedom from discomfort, and the clam is really lower than most worms.

If an animal is to progress, it must keep such a conformity ever secondary to a still more important element, namely, conformity or obedience to the laws of its own structure and being. This second element the mollusk and every creeping stage neglected, and the result of this neglect was stagnation or degeneration. Activity was essential to progress from the very structure and laws of development of the animal, while a great abundance of food was not. A life of ease, for the same reason, necessarily results in degeneration.

But you will ask, What becomes of Mr. Darwin's theory of evolution, if obedience to the laws of individual being is more important than conformity to external conditions? Both are evidently necessary, and they are not so different as they may seem at first sight. They are really one and the same. Bringing out the best and highest there is in us, is the only true conformity to that which is deepest and surest and most enduring in our environment. That in environment which makes for digestion is almost palpable and tangible, that which makes for activity less so perhaps; but that which makes for brain and truth and right is intangible and invisible. We easily fail to notice it; and, unless we take a careful view of the course of development in the highest forms of life, we may be inclined to deny its existence. But it is surely there, if man is a product of evolution.

Each successive stage of animal life is not the preceding stage on a higher plane, but the preceding stage modified in conformity to the environment of that from which it has just arisen. Says Professor Hertwig * : " During the process of organic development the external is continually becoming an integral part of the individual. The germ is continually growing and changing at the expense of surrounding conditions." Every stage thus contains the result of a host of reactions to a ruder and older portion of environment. And the higher we go the more has the original protoplasm and structure been modified as the result of these reactions.

We have seen clearly that environment must be studied through its effect upon living beings. Viewed

---

* Hertwig : Zeit- und Streitfragen, p. 82.

from any other standpoint it appears to be a myriad, almost a chaos, of interacting, apparently conflicting, forces. The resultant of some of these is shown by the animal at any stage of its development. And as the animal advances, the resultant determining its new line, or stage, of advance, includes new forces, to which it has only lately become sensitive. And thus the human mind, as the last and highest product of evolution, mirrors most adequately the resultant of all its forces. If we would know environment we must study ourselves, not atoms alone, nor rocks, nor worms.

Extremely sensitive photographic plates, after long exposure, have proven the existence of stars so dim and far-off as to be invisible to the best telescopes. Man's mind is just such a sensitive plate ; it is the only valid representation of environment.

The truth would appear to be that the law is present in environment, but hard to read ; but it is stamped upon our structure and being so deeply and plainly that the dullest of us cannot fail to read it. We learned the fact of gravitation the first time that we fell down in learning to walk, long afterward we learned that its law guided earth and moon. And it is the presence of this law within us, and our own knowledge that we are conscious of it, that makes man without excuse. But conformity to that which is deepest in environment often, always, demands non-conformity to some of the most palpable of surrounding conditions.

There is no better statement of the ultimate law of conformity than the words of Paul : " Be not conformed to this world ; but be ye transformed by the renewing of your mind, that ye may prove what is that good and acceptable and perfect will of God."

And this difference is exactly what I have been trying to put before you. The mollusk conformed, but the vertebrate conformed in a very different way, and was transformed, "metamorphosed," to translate the Greek word literally, into something higher. And let us not forget that man conforms consciously and voluntarily, if at all; he is able to read in himself and environment the law to which lower forms have been compelled unconsciously to conform.

These facts merely illustrate a great law of life. No man's eye, much less hand, can grasp the whole of the present and at the same time the future. Rather what we usually call present advantage is not advantage at all, but the first step in degeneration. If one will be rich in old age he must deny himself some gratifications in youth; his present reward is his self-control. If a man will climb higher than his fellows he must expect to be sometimes solitary; his reward is the ever-widening view, though the path be rougher and the air more biting than in their lower altitude. If he point to heights yet to attain, the majority will disbelieve him or say, " Our present height was good enough for our ancestors, it is good enough for us. Why sacrifice a good thing and make yourself ridiculous scrambling after what in the end may prove unattainable?" If you discover new truths you will certainly be called a subverter of old ones. And this is entirely natural. The upward path was never intended to be easy.

Read the " Gorgias " of Plato, and let us listen to the closing words of Socrates in that dialogue: " And so, bidding farewell to those things which most men account honors, and looking onward to the truth, I shall earnestly endeavor to grow, so far as may be, in goodness, and thus live, and thus, when the time comes, die. And, to the best of my power, I exhort all other men also; and you especially, in my turn, I exhort to this life and contest, which is, I protest, far above all contests here." You must remember that Callicles has been taunting Socrates with his lack of worldly wisdom and the certainty that in any court of justice he would be absolutely helpless because of his lack of knowledge of the rhetorician's art : " This way then we will follow, and we will call upon all other men to do the same, not that which you believe in and call upon me to follow; for that way, Callicles, is worth nothing."

And Socrates met the end which he expected : death at the hands of his fellow-citizens.

And here perhaps a little glimmer of light is thrown into one of the darkest corners of human experience. The wise old author of Ecclesiastes writes : " There is a just man that perisheth in his righteousness ; and there is a wicked man that prolongeth his life in his wickedness. There is a vanity which is done upon the earth, that there be just men unto whom it happeneth according to the work of the wicked ; again, there be wicked men to whom it happeneth according to the work of the righteous : I said that this also is vanity." "I returned and saw under the sun that the race is not to the swift, nor the battle to the strong, neither yet bread to the wise, nor yet riches to men of understanding, nor yet favor to men of skill ; but time and chance happeneth to them all " (Eccles. viii. 14. ;

202 THE WHENCE AND THE WHITHER OF MAN

ix. 11). It is this element of chance that threatens to
make a mockery of effort, and sometimes seems to
make life but a travesty. The terrible feature of Ten-
nyson's description of Arthur's last, dim battle in the
west is not the "crash of battle-axe on shattered helm,"
but the all-engulfing mist.

Perhaps this is all intended to teach us that riches
and favor, and even bread, are not the essentials of life,
and that failure to attain these is not such ruin as we
often think. But no man ever struggled for wisdom,
righteousness, unselfishness, and heroism without attaining them ; even though the more he attained the
more dissatisfied he became with all previous attainment. And if our slight attainments in wisdom and
knowledge always brought wealth and favor, we might
rest satisfied with the latter, instead of clearly recognizing that wisdom must be its own reward. Uncertainty and deprivation are the best and only training for a hero, not sure reward paid in popular
plaudits.

Political economists speak of the productiveness and
prospectiveness of capital. We may well borrow these
terms, using them in a somewhat modified sense. In
our sense capital is productive in so far as it gives
an immediate return; it is prospective in proportion
as the return is expected largely in the future. A
"pocket" may yield an immediate very large return of
gold nuggets at a very slight expense of labor and ap-
pliances, but it is soon exhausted. In a mine the ore
may be poor near the surface, but grow richer as the
shaft deepens ; the vein is narrow above, but widens
below. The returns are at first small, its inexhaustible
richness becomes apparent only after considerable

time and labor. The value of the "pocket" is purely productive, that of the mine largely or purely prospective. Indeed it may be opened at a loss. But even a rich mine may be worked purely for its productive value ; it may be " skinned."

Let us apply this thought to the development of a species ; although what is true of the species will generally be true of the individual also, for the development of the two is, in the main, parallel. In the animal all functions are to a certain extent productive, and all directly or indirectly prospective. When we examine the sequence of functions we cannot but notice how largely their value is prospective. As long as a lower function is rising to supremacy in the animal, it appears to be retained purely for its productive value; thus digestion in hydra or gastræa. But after a time animals appeared which had some muscle and nerve. And, by the process of natural selection, those animals which used digestion as an end for its productive value became food for, and gave place to, those using it as a means of supporting muscle and nerve of greater prospective value. And similarly, those animals which used muscle, or even mind, productively gave place to others using these prospectively.

In other words, the functions and capacities of any animal, the extent of its conformity to environment, may be regarded as its capital. The animal may use this capital productively or prospectively. It may spend its income, and more too ; it may increase its capital. Now social capital will always fall sooner or later to those communities whose members use it most prospectively, who are willing to forego, to quite an

extent, present enjoyment, and look for future return. The same is true of all development. Sessile
forms and mollusks, and, in a less degree, crabs and
reptiles, worked for immediate return. They are like
extravagant heirs who draw on their capital and
sooner or later come to poverty. The primitive vertebrate, the mammal, and the other ancestors of man
used their capital prospectively, and it increased, as if
at compound interest.

The spendthrift appears at first sight to have the
greatest enjoyment in life, the rising business man
works hard and foregoes much. I believe that the
latter is really by far the happier of the two. But, if
you can spend only a day or two in a city, and your
examination is superficial, you may easily make the
mistake of considering the spendthrift as the most
successful man in the community. So, in our brief
visit to the world in times past, we picked out the
crab, the reptile, and the carnivore as its rising members.

Once more, capital can be spent very quickly; to
use it prospectively requires time. This is a truism ;
but it does no harm to call attention to truisms which
have been neglected. Organs and powers of great
prospective value are slow and difficult of development. If their increase is to be at all rapid, they
must start early. If their development and culture
is deferred, there will be little or no advance, but
probably degeneration. Extravagance grows rapidly
and soon becomes irresistible ; habits of saving must
be formed early. The same is true of the development of all other virtues.

There is in the child an orderly sequence of development of mental traits. While these powers are in their earlier, so to speak embryonic, stages of development, they can be fostered and increased or retarded. They are still plastic. Very early in a child's life acquisitiveness shows itself; he begins to say " I," and "mine," and desires things to be his "very own." And this can be fostered so that the child will grow up a "covetous machine." Or he may be taught to share with others.

Not so much later, while the child is still in the lower grades of his school life, comes the period of moral development. If, during this period, these powers are fostered and cultivated, they may, and probably will, be dominant throughout his life. And herein lies the dignity and glory of the unappreciated, underpaid, and overworked teachers of our "lower" schools, that they have the opportunity to cultivate these moral powers of the child during these most critical years of his life. Repression or neglect here works life-long and irreparable harm. The young man goes out into the world. Here "practical" men continually instruct him by precept upon precept, line upon line, that he cannot afford to be generous until he has acquired wealth; that he must first win success for himself, and that he can then help others. And, unless his character is like pasture-grown oak, he follows and improves upon their teachings. *He reverses the sequence of functions.* He puts acquisitiveness first and right and sterling honesty and unselfishness second. For a score or more of years he labors. At first he honestly intends to build up a strong character and a generous nature just as soon as he can afford to; but for the present he

cannot afford it. If he is to succeed, he must do
as others do and walk in the beaten track. He wins
wealth and position, or learning and fame. He now
has the ability and means to help others, but he no
longer cares to do so. Loyalty to truth, sterling honesty—the genuine, not the conventional couuterfeit—
unselfishness, in one word, character, these are plants
of slow growth. They require cultivation by habit
through long years. In his case they have become
aborted and incapable of rejuvenescence. But his
rudiment of a moral nature feels twinges of remorse.
He ought not to have reversed the sequence of functions, and he knows it. But he cannot retrace his
steps. He made the development of character impossible when he made wealth his first and chief aim.
If he has a million dollars he tries to insure his soul
by leaving in his will one-tenth to build a church, or,
possibly, one-half for foreign missions. In the latter
case he will be held up as a shining example to all
the youth of the land, and the churches will ring with
his praises. But what has been the effect of his life
on the moral, social capital of the community? Is the
world better or worse for his life? He has all his
life been disseminating the germs of · a soul-blight
more infectious and deadly than any bodily disease.

If he has made learning or fame his chief aim, he
probably has not the money to buy soul-insurance.
He takes refuge in agnosticism, like an ostrich in a
bush. His agnosticism is in his will; he does not
wish to see. Or its cause is atrophy, through disuse, of
moral vision. He cannot see. There are agnostics of
quite another stamp, whom we must respect and honor
for their sterling honesty and high character, though

we may have little respect for their philosophical
tenets. But how much has our scholar advanced the
morality of the community? He has probably done
even more harm than the business man, who is a mere
" covetous machine."

The " practical " man has reversed the sequence of
functions. Character is, and must be, first ; and
wealth, learning, power, and fame are the materials,
often exceedingly refractory, which it must subjugate
to its growth and use. And this subjugation is anything but easy. The reversal of the sequence results
in a moral degradation and poverty indefinitely more
dangerous to the community than the slums of our
great cities. For these may be controlled and
cleansed; but the moral slum floods our legislatures
and positions of honor and trust, and invades the
churches. The mental and moral water-supply of the
community is loaded with disease-germs.

The social wealth of a community is the sum total
of the wealth of its individual members. And a community is truly wealthy only when this wealth is, to a
certain extent, diffused. If there is any truth in our
argument that the sequence of functions culminates
in righteousness and unselfishness, the real social
wealth of a community consists in its moral character,
not in its money, or even in its intelligence. We may
rest assured that character, resulting in industry and
economy, will bring sufficient means of subsistence,
so that all its members will be fed and housed and
clothed. And art and culture, of the most ennobling
and inspiring sort, will surely follow. And even if
such literature failed as largely composes our present
fin-de-siècle garbage-heap, we would not regret its ab-

sence. That community will and must survive in which the largest proportion of members make the accumulation of character their chief and first aim. And to this community every rival must in time yield its place and power, and all its acquisitions. And in every advancing community the position of any class or profession will in time be determined by its moral wealth.

But this moral wealth is intangible. The rewards and penalties of moral law easily escape notice in our hasty and superficial study of life. The God immanent in our environment often seems to hide himself. The altar of Jehovah is fallen down, and Baal's temples are crowded with loud-mouthed worshippers. The bribes of present enjoyment and of immediate success loom up before us, and we doubt if any other success is possible.

But the law of progress, even now so dimly discernible in environment, is written in our minds in letters of fire. For we have already seen that environment can be understood only by tracing its effects in the development of life. What is best and highest in us is the record of the working of what is best and highest in environment. And the personal God so dimly seen in environment is revealed in man's soul. Man must study himself, if he is to know what environment requires of him. And if the knowledge of himself and of the laws of his being is the highest knowledge, is not the vision of, and struggle toward, higher attainments, not yet realized and hence necessarily foreseen, the only mode of farther progress? And what is this pursuit of, and devotion to, ideals not yet realized and but dimly foreseen, if it is not

Faith, "the substance of things hoped for, and evidence of things not seen?" By it alone can man "obtain a good report." Man must "walk by faith, not by sight." "For the things which are seen are temporal, but the things which are not seen are eternal."

14

CHAPTER VIII

IN Kingsley's fascinating historical romance, Raphael Abeu-Ezra says to Hypatia, "Is it not possible that we have been so busy discussing what the philosopher should be, that we have forgotten that he must first of all be a man?" This truth we too often forget. No statesman, philosopher, least of all teacher, can be truly great who is not, first of all, and above all, a great man. And in our study of man are we not prone to forget that he stands in certain very definite and close relations with surrounding nature?

Man has been the object of so much special study, his position, owing to his higher moral and mental power, is so unique that he has often been regarded not only as a special creation, but as created to occupy a position not only unique, but also exceptional, above many of the very laws of nature, and not bound by them. Many speak and write of him as if it were his chief glory and prerogative to be as far removed as possible, not only from the animal, but even from the whole realm of nature. The mistake of making him an exception arises, after all, not so much from too high a conception of man, at least of his possibilities, as from too low a view of nature.

But however this view may have arisen, it is one-sided and mistaken. Man certainly has a place in

Nature—not above it. If he is the goal toward which
the ascending series of living forms has continually
tended, he is a part of the series—the real goal lies far
above him.

Pascal says, " It is dangerous to show a man too
clearly how closely he resembles the brute without
showing him at the same time his greatness. It is
equally dangerous to impress upon him his greatness
without his lowliness. It is still more dangerous to
leave him in ignorance of both. But it is of great advantage to point out to him both characteristics side
by side."

A great German thinker began his work on the
human soul with a discussion of the law of gravitation.

All study of man must begin with the study of the
atom. Man's life we have seen to be the aggregate of
the work of all the cells of his body. But the protoplasm which composes his cells is a chemical compound, and hence subject to all the laws of all the
atoms of which it is composed. And its molecules,
or the smallest mechanically separable compounds of
these atoms, are arranged and related according to the
laws of physics, so as to permit or produce the play
of certain forces which are always the result of atomic
or molecular combination. Every motive or thought
demands the combustion of a certain amount of
material which has been already assimilated in the
microscopic cellular laboratories of our body. Every
vital activity is manifested at least through chemical
and physical forces. And the elements of the fuel
for our engines we receive through plants from the
inorganic world. For the plant, as we have seen,
stores up as potential energy in its compounds the

actual energy of the sun's rays. And thus man lives and thinks by energy, obtained originally from the sun. But man not only consumes food and fuel. The complicated protoplasm is continually wearing out and being replaced. Every cell in our bodies is a centre toward which particles of material stream to be assimilated and form for a time a part of the living substance, and then to be cast out again as dead matter. Our very existence depends upon this continual change. There is synthesis of simple substances into more complex compounds, and then analysis of these complex compounds into simpler, and from this latter process results the energy manifested in every vital action. We are all whirlpools on the surface of nature; when the whirling ceases we disappear. Man, like every other living being, exists in a condition of constant interchange with surrounding nature; he is rooted in innumerable ways in the inorganic world.

And because of these close relations the great characteristic of living beings is the necessity and power of conformity to environment. Hence a very common definition of life is the continual adjustment of internal relations to external relations or conditions. To a very slight extent man can rise superior to certain of the ruder elements of his surroundings, but he gains this victory only by learning and following the laws of the very environment which he succeeds in subjecting to himself. Indeed his higher development and finer build bring him into touch with an indefinitely wider range of surroundings than even the lower animal. Forces, conditions, and relations which never enter the sphere of life of lower forms, crowd and press upon him and he cannot escape them. His higher position, instead of freeing him from dependence upon environment and subjection to law, makes him thus more sensitive, as well as more capable of exact conformity to an environment of almost infinite complexity; and more sure of absolute ruin, if ignorant, negligent, or disobedient. The words of the German poet are literally true :

> " Nach ehernen, eisernen, grossen Gesetzen,
> Müssen wir alle unseres Daseins
> Kreise vollenden."

But man is an animal. And the principal characteristic of an animal is that it eats a certain amount of solid food. The plant lives on fluid nutriment, and this comes to it by the process of diffusion in every drop of water and breath of air. The acquisition of food requires no effort, and the plant makes none. It has therefore always remained stationary and almost insensible. Not taking the first step it has never taken any of the higher ones. But solid food would not, as a rule, come to the animal—though stationary and sessile animals are not uncommon in the water— he must go in search of it. This called into play the powers of locomotion and perception. And in the sequence of function we have seen digestion calling for the development of muscle; and muscle, of nerve and brain. And the brain became the organ of mind.

Man as a mere animal is necessarily active and energetic; otherwise he stagnates and degenerates. Labor is a curse, but work a blessing; and man's best work, of every kind, is done in the friction of life, not in ease and quiet. Man is, further, a being composed of cells, tissues, and organs, which were successively developed for him by the lower animal kingdoms. The old view, that man was the microcosm, had in it a certain amount of every important truth. We need to be continually reminded of our indebtedness in a thousand ways to the lowest and most insignificant forms of life.

Man is a vertebrate animal. This means that he has a locomotive, not protective, skeleton, composed of cartilage—a tough, elastic, organic material, hardened, as a rule, by the deposition of mineral salts, mainly phosphate of lime, in exceedingly fine particles, so as to form a homogeneous, flawless, elastic, tough, light, and unyielding skeleton, held together by firm ligaments.

The skeleton is internal, and this fact, as we have seen, gives the possibility of large size. And size is in itself no unimportant factor. Professor Lotze maintains that without man's size and strength, agriculture and the working of metals, and thus all civilization, would have been impossible. But we have already seen that there is an extreme of size, *e.g.*, in the elephant, which makes its possessor clumsy, able to exist only where there are large amounts of food in limited areas, slow to reproduce, and lacking in adaptability. This extreme also is avoided in man ; in this, as in many other particulars, he holds the golden mean. But we have also seen that large size is, as a rule, correlated with long life and great opportunity for experience and observation. And these are the foundations of intelligence. Hence the deliverance of the higher vertebrate, and especially of man, from any iron-bound subjection to instinct.

And here another question of vital importance

meets us. Is man's life at present as long as it should or can be? The question is exceedingly difficult, but a negative answer seems more probable. We cannot but hope that, with a better knowledge of our physical structure, a clearer vision of the dangers to which we are exposed, more study of the laws of physiology, heredity, and of our environment, and above all, less reckless disregard of these in a mad pursuit of pleasure, wealth, and position, man's period of mature, healthy, and best activity may be lengthened, perhaps, even a score of years. The mitigation of hurry and worry alone, the two great curses of our American civilization, might postpone the collapse of our nervous systems longer than we even dream. And if we could add even five years to the working life of our statesmen, scholars, and discoverers, the work of these last five years, with the advantage of all previously acquired knowledge and experience, might be of more value than that of their whole previous life. Human advance could not but be greatly, or even vastly, accelerated.

Moreover, we have seen that the history of vertebrates is really the history of the development of the cerebrum, forebrain or large brain, as we call it in man. This is the seat in man of consciousness, thought, and will. This portion as a distinct and new lobe first appears in lowest vertebrates, increases steadily in size from class to class, reaches its most rapid development by mammals, and its culmination in man. During the tertiary period—the last of the great geological periods—the brain in many groups of mammals increased in size, both absolutely and relatively, eight to tenfold. Dr. Holmes says, that the education of a child should begin a century or two before its birth; man really began his mental education at least as early as the appearance of vertebrate life.

But man is a mammal. This means that every organ is at its best. The digestive system, while making but a small part of the weight of the body, and built mainly on the old plan, is wonderfully perfect in its microscopic details. The muscles are heavy and powerful, arranged with the weight near the axis of the body, and replaced near the ends of the appendages by light, tough sinews. The higher mammal is this compact, light, and agile. The skeleton is strong, and the levers of the appendages are fitted to give rapidity of motion even at the expense of strength. And this again is possible only because of the high development and strength of the muscles. Moreover, the highest mammals are largely arboreal, and in connection with this habit have changed the foreleg into an arm and hand. The latter became the servant of the brain and gave the possibility of using tools.

But increase in size and activity, and the expense of producing each new individual, led to the adoption of placental development. And the mammal is so complex, the road from the egg to the fully developed young is so long, that a long period of gestation is necessary. And even at birth the brain, especially of man, is anything but complete. Hence the necessity of the mammalian habit of suckling and caring for the young. And this feebleness and dependence of the young had begun far below man to draw out maternal tenderness and affection. And the mammalian mode of reproduction and care of young led to a more marked difference and interdependence between the sexes.

The result of this is man's family life, as Mr. John Fiske has shown so beautifully in that fascinating monograph, " The Destiny of Man." And family life once introduced becomes the foundation and bulwark of all civilization, morality, and religion. Far down in the mammalian series, before the development of the family, maternal education has become prominent, and the young begins life, benefited by the experiences of the parent. How much more efficient is this in family life. But, furthermore, the family is perhaps the first, certainly the most important, of those higher unities in which men are bound together. Social life of a sort undoubtedly existed, before man, among birds, insects, and lower mammals. The community was often defective or incomplete in unity, or existed under such limitations that it could not show its best results, but that it was of vast benefit from an even higher than mere physical standpoint, no one will, I think, deny. But with the family a new era of education and social life began.

First of all, the struggle for existence is thereby greatly modified and mitigated. This crowding out and trampling down of the weaker by the stronger is transferred, to a certain extent, from the individual to the family and, in great degree, from the family to larger and larger social units. For within the limits of the family competition tends to be replaced by mutual helpfulness, and not only are the loneliness and horror of the struggle between isolated individuals banished, but, what is vastly more, the family becomes the school of unselfishness and love. And what has thus become true of the single family, and groups of nearly related families, is slowly being realized in the larger units of communities and states. For, as families and communities are just as really organisms as are the individual men and women, whose soundness depends upon the healthy activity of every organ, so there is a survival, first of families, then of communities and rival civilizations, in proportion to their unity and soundness in every part. For on account of the close bonds of family and social life, and in connection with the development of articulate speech, a new kind of heredity, so to speak, arises, of vast importance for both good and evil. This mental and moral heredity, overleaping all boundaries of blood and natural kinship, spreads light and good influence or an immoral contagion through the community. And thus, in sheer self-defence, society passes laws setting limits to the oppression of the poor and weak, lest, degraded and brutalized, they become breeding centres of physical and moral disease in the community. The positive lesson that the surest mode of self-defence is the elevation of these submerged classes, we are just beginning to learn and apply.

By the ever-increasing acceleration of the development the gap between man and the lower animal widens with wonderful rapidity. Of course it is only in man, and higher man, that these last and highest results of mammalian structure appear. But that, far removed as they are, they are the results of mammalian and vertebrate characteristics cannot, I think, be well denied. And this is only one of innumerably possible illustrations of the fact that all our most highly prized institutions are rooted far back in our ancestry, often ineradicably in the very organs of our bodies. And thus evolution, which many view only from its radical

side—and it has a radical side—is really the conservative bulwark of all that is essentially worth possessing
in the past.

But every factor in man's development tends toward
intellectual and spiritual development. Man's vast
increase of brain; his finely balanced body; his upright
gait; setting his hands free from the work of locomotion that they might become the skilful servants of
the mind; finally, articulate speech and social, and,
above all, family, life, all tended in this same direction.

And this makes the great difficulty in assigning man
his proper place in our systems of classification. Our
zoölogical classifications depend upon anatomical characteristics; and anatomically man belongs among the
order primates. But mental and moral values cannot
be expressed in terms of anatomy, any more than we
can speak of an idea of so many horse-power, and
hence worth three or four ancestral dollars. Hence,
while from the zoölogical standpoint man is a primate,
and while he is very probably descended from one of
these, he has gradually risen above them mentally and
spiritually, so that he stands as far above them as they
above the lowest worm. And this leads us to the consideration of man, not merely as a mammal, but as
" Anthropos," Homo sapiens, although he often degenerates into "Simia destructor."

From what has just been said man's pre-eminence
cannot consist in any anatomical characteristic, even
of the brain—much less of thumb, forefinger, hand, or
foot. But man's mental and moral characteristics
(even though germs of these may be present in the
animal), whether differing in degree or kind from theirs,
raise his life to a totally different plane. He lives in

an environment of which the lower animal is as unconscious and ignorant as we of a fourth dimension of
space. He has the knowledge of abstract truth and
goodness, of certain standards outside of mere appetite
and desire, and feels and acknowledges, however dimly,
the requirement and the ability to conform his life to
these standards. He alone can say " I ought," and
answer "I can and will." And hence man alone actually
lives in an environment of the laws of reason, responsibility, and personality. Whatever germs of these
higher powers the animal possesses are means to material ends, to the physical life of the animal. In man the
long and slow evolution has ended in revolution, the
material and physical have been dethroned, and truth
and goodness reign supreme as ends in themselves.

But, you may object, this definition of man may be
true ideally, certainly it is not true actually. Where
are the high ideals of truth and goodness in the savage ? and are these the supreme ends of even the
average American of to-day? But allowing all weight
to this objection, does it not remain true that a being
who never says " I ought," who acknowledges and
manifests no responsibility, to whom goodness does not
appeal, and in whom these feelings cannot be awakened, is either not yet or no longer man ? But far
more than this, if the character of the individual is to
be judged by his tendency more than his present condition, by the way in which he is going more than his
momentary position, is not the race to be judged and
defined by a tendency, gradually though very slowly
becoming realized, and a goal, toward which it looks
and which it is surely attaining, rather than by its

present realization? As we rise higher in the animal kingdom the characteristics of the successive higher groups are more and more slow of attainment and difficult of realization, just because of their grander possibilities. And this is true and important above all in the case of man. His possibilities are beyond our powers of conception, for, if you will, man is yet only larval man.

We have followed the sequence of functions to its culmination in a mind completely dominated by righteousness and unselfishness, however far above our present attainments this goal may be. We have found that all attempts to reverse this sequence end in death or degeneration. Failure to advance, especially in higher forms, results in extinction or retrogression. We cannot stand still. Each higher step is longer and more important than any preceding; each last step is essential to life. Righteousness in the will is the last step essential to man's progress. And if a sound mind in a sound body is important or necessary, a sound will, resolutely set on right, is absolutely essential. Failure to attain this is ruin.

And man can to a great extent place himself so that his surroundings shall aid him to take this last, essential, upward step. He does this by the choice of his associates. If he associates himself with men who are tending upward, he will rise ever higher. If he choose the opposite kind of associates he must sink into ever deeper degradation; he has thereby chosen death. For his associates, once chosen, make him like themselves. And thus natural selection makes for the survival of those men who resolutely choose life. And thoughtless or careless failure to choose is ruin. The man has preferred degradation ; it is only right that he should have it to satiety.

But man is not, and never can be, pure spirit. He may " let the ape and tiger die," but he must always retain the animal with its natural appetites. Moreover, his higher mental capacities increase their power. Memory recalls past gratifications as it never does to the animal ; imagination paints before him vivid pictures of similar future enjoyments, and mental keenness and strength of will tell him that they can all be his. But if he yields himself a slave to these appetites, if he seeks to be an animal rather than a spiritual being, he becomes not an animal but a brute ; and the only genuine brute is a degenerate man. And thus after conquering the world man's very structure compels him to join battle with himself. For here, as everywhere else, to attempt to go backward to a plane of life once passed is to surely degenerate. The time when the prize of pre-eminence could be won by mere physical superiority was passed before man had a history. Physical superiority must be maintained, and every advance in art and science, considered here as ministering to man's physical comfort, is advantageous just so far as these allow man freedom and aid to pursue the mental and moral line which is the only true path left open to him. But when even these are allowed to minister only to the animal, or to tempt to luxurious ease and indifference to any higher aims, in a word, in so far as they fail to minister to mental and moral advancement, they are in great danger of becoming, if they have not already become, a curse rather than a blessing. And we all know that this has been proven over and over again in human history.   Families, cities, and nations rot, mainly because they cannot resist the seductions of an overwhelming material prosperity. A man says to his soul, " Take thine ease, eat, drink and be merry," and to that man scripture and science say, with equal emphasis, " Thou fool ! "

Every upward step in attainment of the comforts of life, of art and science, brings man into new fields not of careless enjoyment but of struggle.   They swarm with new enemies and temptations before unknown. The new attainments are not unalloyed blessings, they are merely opportunities for victory or defeat. The uncertain battle is only shifted to a little higher plane.   Man has increased the forces at his command only to meet stronger opposing hosts.   And retreat is impossible.   Man remains a spiritual being only on condition that he resolutely and vigilantly purposes to be so.   To lag behind in this spiritual path is death.

And the epitaph of nations and individuals is the record of their defeat in this struggle to be masters and not slaves of their material and intellectual attainments.   Greece, the most intellectual of all nations of all times, died in mental senility of moral paralysis.   Of Socrates's and Plato's " following after truth " nothing remained but the gossipy curiosity of a second childhood, living only to tell or to hear some new thing.   And the schools of philosophy were closed because they had nothing to tell which was worth the knowing or hearing.   All the wealth of the world was poured into Rome, the home of Stoic philosophy, and it was smothered, and died in rottenness under its material prosperity.

A family, race, or nation starts out fresh in its youthful physical and mental vigor and strict obedience to moral law and in its faith in God. For these reasons it survives in the struggle for existence. It grows in extent and power, in intelligence and wealth. But with this increase in wealth and power comes a deadening of the mind to the claims of moral law, and an idolatrous worship of material prosperity. The new generation looks upon the stern morality and industry and self-control of its ancestors as straight-laced and narrow. Morality may not be unfashionable, but any stern rebuke of immorality is not conventional. Strong moral earnestness and whole-souled loyalty to truth are not in good form. Wealth and social position become the chief ends of men's efforts, and, to buy these, unselfishness and truth and self-respect are bartered away. Luxury, enervation, and effeminacy are rife, and snobbery follows close behind them. The ancestral vigor, the insight to recognize great moral principles, and the power to gladly hazard all in their defence have disappeared in a mist of indifference, which beclouds the eyes and benumbs all the powers. The race of giants is dwindling into dwarfs. They say, when the time comes, we will rouse ourselves and be like our fathers. And the crisis comes, but they are not equal to it. The nation has long enough cumbered the ground, it has already died by suicide and must now give place to a race and civilization which has some aim in, and hence right to, existence, and which is of some use to itself and others. If we would learn by observation, and not by sad experience, we must remember that man is above all, and must be a religious being conforming to the personality of the God manifested in his environment.

Can you find anywhere a more profound or scientific philosophy of history than that of Paul in the first chapter of Romans? "For the invisible things of him since the creation of the world are clearly seen, being perceived through the things that are made, even his everlasting power and divinity; so that they are without excuse: because that, knowing God, they glorified him not as God, neither gave thanks; but became vain in their reasonings and their senseless heart was darkened. Professing themselves to be wise, they became fools. And even as they refused to have God in their knowledge, God gave them up to a reprobate mind, to do those things which are not fitting; being filled with all unrighteousness." * And then follows the dark picture, from which we revolt but which the ancient historians themselves justify.

On the ceiling of the Sistine Chapel at Rome is Michel Angelo's marvellous painting of the creation of Adam. A human figure of magnificent strength is half-rising from its recumbent posture, as if just awakening to consciousness, and is reaching out its hand to touch the outstretched finger of God. The human being became and becomes man when, and in proportion as, he puts himself in touch with God, and is inspired with the divine life. The lower animal conformed mainly to the material in environment, man conforms consciously to the spiritual and personal.

Any science of human history that does not acknowledge man's relation to a personal God is fatally

* Romans i. 20-22, 28.

15

incomplete; for it has missed the goal of man's development and the chief means of his farther advance. And a religion which does not emphasize this is worse than a broken reed. It is a mirage of the desert, toward which thirsty souls run only to die unsatisfied.

Man can never overcome in this battle with the allurements of material prosperity and with the pride and selfishness of intellect, except as he is interpenetrated and permeated with God, any more than we can move or think, unless our blood is charged with the oxygen of the air. It is not enough that man have God in his intellectual creed; he must have him in his heart and will, in every fibre of his personality, in every thought and action of life. Otherwise his defeat and ruin are sure.

Three fatal heresies are abroad to-day: 1. Man's chief end is avoidance of pain and discomfort, in one word, happiness; and God is somehow bound to surfeit man with this. And this is the chief end of a mollusk. 2. Man's chief end is material prosperity and social position. 3. Man's chief end is intellect, knowledge. Each one of these three ends, while good in a subordinate place, will surely ruin man if made his chief end. For they leave out of account conformity to environment. "Man's chief end is to glorify God and enjoy him for ever." And just as the plant glorifies the sun by turning to, and being permeated and vivified and built up by, the warmth and light of its rays, similarly man must glorify God. This is the religion of conformity to environment: man working out his salvation because God works in him. Thus, and thus only, shall man overcome the allurements of these lower endowments and receive the rewards of "him that overcometh."

Thus prosperity and adversity, success and failure, continually test a man. If he can rise superior to these, can subjugate them and make them subserve his moral progress, he survives ; if he is mastered by them, he perishes. Through these does natural selection mainly work to find and train great souls. They are the threads of the sieve of destiny.

In this struggle man must fight against overwhelming odds, and the cost of victory is dear. He must be prepared, like Socrates, to "bid farewell to those things which most men count honors, and look onward to the truth." He appears to the world at large, often to himself, eminently unpractical. The majority against his view and vote will usually be overwhelming. Truth is a stern goddess, and she will often bid him draw sword and stand against his nearest and dearest friends. The issue will often appear to him exceeding doubtful. The grander the truth for which he is fighting, the greater the need of its defence and enforcement, the greater the probability that he will never live to see its triumph. The hero must be a man of gigantic faith. But all his ancestors have had to make a similar choice and to fight a similar battle. The upward path was intended to be exceedingly hard. This is a law of biology.

Why this is so I may not know. I only know that no better and surer way could have been discovered to train a race of heroes. For no man ever becomes a hero who has not learned to battle with the world and himself. Does it not look as if God loved a heroic soul as much as men worship one, and as if he

intended that man should attain to it? Man was
born and bred in hardship that he might be a hero.

> "Careless seems the great avenger; history's pages but record
>   One death-grapple in the darkness 'twixt old systems and
>   the word;
>   Truth forever on the scaffold, Wrong forever on the throne,
>   Yet that scaffold sways the future, and behind the dim un-
>   known
>   Standeth God within the shadow, keeping watch above his
>   own.
>
> "Then to side with Truth is noble when we share her wretched
>   crust,
>   Ere her cause bring fame and profit, and 'tis prosperous to
>   be just;
>   Then it is the brave man chooses, while the coward stands
>   aside,
>   Doubting in his abject spirit, till his Lord is crucified,
>   And the multitude make virtue of the faith they had de-
>   nied."

The Crown Prince of Prussia has less spending
money than many a young fellow in Berlin. He is
trained to economy, industry, self-control. He is to
learn something better than habits of luxury, to rule
himself, and thus later the German Empire. The
children of a great captain, themselves to be soldiers,
must endure hardness like good soldiers. And man
is to fight his way to a throne.

But his powers are still in their infancy and the
goal far above him. What he is to become you and I
can hardly appreciate. First of all, the body will become finer, fitted for nobler ends. It will not be allowed to degenerate. It may become less fitted for
the rough work, which can be done by machinery; it

will be all the better for higher uses. It is to be transformed, transfigured. The eye may not see so far, it will be better fitted for perceiving all the beauties of art and nature. It will become a better means of expressing personality, as our personality becomes more "fit to be seen." It is continually gaining a speech of its own. And will not the ear become more delicate, a better instrument for responding to the finest harmonies, and better gateway to our highest feelings? We may not have so many molar teeth for chewing food, but may not our mouths become ever finer instruments for speech and song? In other words, the body is to be transfigured by the mind and become its worthy servant and representative.

As we learn to live for something better than food and clothes, and cease to pamper the body, it will become better and healthier. Science will stamp out many diseases, and we shall learn to prevent others by right living. And what a change in our moral and religious life will be made by good health. What a cheerful courage and hope it will give.

Man will become more intelligent. He will learn the laws of heredity and of life in general. He will see deeper into the relations of things. He will recognize in himself and his environment the laws of progress. He will clearly discern great moral truths, where we but dimly see lights and shadows.

But while we would not underestimate the value and necessity of growth in knowledge, we must as clearly recognize that the intellect is not the centre and essence of man's being. Knowledge, while the surest form of wealth of which no one can rob us, and the best as the stepping-stone to the highest well-

230 THE WHENCE AND THE WHITHER OF MAN

being, is like wealth in one respect: it is not character and can be used for good or evil. If my neighbor uses his greater knowledge as a means of overreaching us all, it injures us and ruins him.

Our emotions, and this is but another word for our motives, stand far nearer to the centre of life; for they control our conduct and directly determine what we are. Knowledge of environment is good, but of what real and permanent use is such knowledge without conformity? Our real weakness is not our ignorance; we know the good, but lack the will and purpose to live it out. And this is because the thought of truth and goodness excites no such strength of feeling as that of some lower gratification. We cannot perhaps overrate the value of intellect; we certainly underrate the value of emotion and feeling. "Knowledge puffeth up, love buildeth." It does not require great intellect, it does require intense feeling to be a hero. We slander the emotions by calling people emotional because they are always talking about their feelings; but deep feeling is always silent. It is not fashionable to feel deeply, and we are dwarfed by this conventionality. We have almost ceased to wonder, and hence we have almost ceased to learn; for the wise old Greeks knew that wonder is the mother of wisdom.

The man of the future will probably be a man of strong appetites, for he will be healthy; he will be prudent, because wise; but he will hold his appetites well in leash. He will trample upon mere prudential considerations at the call of truth or right. For in him these highest motives will be absolute monarchs, and they are the only motives which can enable a man to face rack and stake without flinching. He will be a hero because he feels intensely. In other words, he will be a man of gigantic will, because he has a great heart. And in the man of the future all these powers will be not only highly developed; they will be rightly proportioned and duly subordinated. He will be a well-balanced man. But how few complete men we now see.

We see the strong will without the clear intellect to guide it; the gush of feeling either directed toward low ends or evaporating in sentiment; the clear head with the cold heart. The high development of one mental power seems to draw away all strength and vitality from the rest. How rarely do we find the strong will guided by the keen intellect toward the highest aims clearly discerned. Memory and imagination must always play their part in the joy set before us. But in addition to all these, the white heat of feeling, of which man alone is capable, is necessary for his grandest efforts. Such a being would be a man born to be a king. And there will be a race of such men. And we must play the man that they may be raised upon our buried shoulders. And they will tower above us, as the seers of old in Judea, Athens, India, and Rome towered above their indolent, luxurious, blind, and material contemporaries. And with all their accelerated development, infinite possibilities will still stretch beyond the reach of their imagination. For "men follow duty, never overtake."

But all our analyses are unsatisfactory. In the history of any great people there is a period when they seem to rise above themselves. They have the strength of giants, and accomplish things before and since impossible. We sometimes ascribe these results to the exuberant vitality of the race at this time; and their life is large and grand. Such was England under Elizabeth. Think of her soldiers and explorers, her statesmen and poets. There were giants in those days. What a healthy, hearty enjoyment they showed in all their work, and with what ease was the impossible accomplished. The greater the hardships to be borne or odds to be faced, the greater the joy in overcoming them. They sailed out to give battle to the superior power of Spain, not at the command, but by the permission, of their queen; often without even this.

And what a vigor and vitality there is in the literature of this period. Life is worth living, and studying, and describing. They see the world directly as it is; not some distorted picture of it, seen by an unhealthy mind and drawn by a feeble hand. The world is ever new and fresh to them because they see it through young, clear eyes.

Were they giants or are we dwarfed? Which of the two lives is normal? They used all their faculties and utilized all their powers. Do we? The only force or product which we are willing to see wasted is the highest mental and moral power. Our engines and turbine wheels utilize the last ounce of pressure of the steam or water. The manufacturers pay high wages to hands who can tend machines run at the highest possible speed. The profits of modern business come largely from the utilization of force or products formerly wasted. But how far do we utilize the highest faculties of the mind, which have to do with character, the crowning glory of human development? Are we not eminently "penny-wise and pound-foolish?" A ship which uses only its donkey-engines, and does nothing but take in and get out cargo is a dismantled hulk. A captain who thinks only of cargo, and engines, and the length of the daily run, but who takes no observations and consults no chart, will make land only to run upon rocks. Are we not too much like such dismantled hulks, or ships sailing with priceless cargoes but with mad captains?

But we have not yet seen the worst results of this waste of our highest powers. The sessile animal, which lives mainly for digestion, does not attain as good digestive organs as his more active neighbor, who subordinates digestion to muscle. Lower powers reach their highest development only in proportion as they are strictly subordinated to higher. This may be called a law of biology. And our lower mental powers fail of their highest development and capacity mainly because of the lack of this subordination.

But a disused organ is very likely to become a seat of disease and to thus enfeeble or destroy the whole body. And this disease effects the most complete ruin when its seat is in the highest organs. Dyspepsia is bad enough, but mania or idiocy is infinitely worse. And our moral powers are always enfeebled, and often diseased, from lack of strong exercise. And some blind guides, seeing only the disease, cry out for the extirpation of the whole faculty, as some physicians are said to propose the removal of the vermiform appendage in children. Similarly might the drunkard argue against the value of brain, because it aches after a debauch. Our work is hard labor, and we gain no enjoyment in the use of our mental powers ; for the enjoyment of any activity is proportional to the height and glory of the purpose for which it is employed. As long as we are content to use only our lower mental faculties and to gain low ends, our use of even these will be feeble and ineffectual, and our lives will be poor, weak, and unhappy.

But future man will subordinate these lower powers to the higher. He will utilize all that there is in him. And his efficiency must be vastly greater than ours. And finally, and most important, these men will be all-powerful, because they have so conformed to environment that all its forces combine to work with them.

England under Elizabeth seemed to rise above itself. Think of Holland, under William the Silent, defying all the power of Spain. Look at Bohemia, under Ziska, a handful of peasants joining battle with and defeating Germany and Austria combined. Think of Cromwell and his Ironsides, before whom Europe trembled. These men were not merely giants, they were heroes. And the essence of heroism is self-forgetfulness. The last thought of William the Silent was not for himself, but for his "poor people." And those rugged Ironsides, "fighting with their hands and praying with their hearts," smote with right good-will and irresistibly, because they struck for truth and freedom, for right and God. These are motives of incalculable strength, and they transfigure a man and raise him above his surroundings and even himself. The man becomes heroic and godlike, and when possessed by these motives he has clasped hands with God. He is inspired and infused with the divine power and life. Such a man has no time nor care to think of himself. To him it matters little whether he lives to see the triumph of his cause, provided he can hasten it. Though victory be in the future, it is sure; and the joy of battle for so sure and grand a triumph is present reward enough. His very faith removes mountains and turns to flight armies of the aliens. For heroism begets faith, just as surely as faith begets heroism.

"Where there is no vision the people perish." When the member of Congress can see nothing higher than spoils of office, nothing larger than a silver dollar, you should not criticise the poor man if his oratorical efforts do not move an audience like the sayings of Webster, Lincoln, or Phillips.

Future man will be heroic and divine, because he will live in an atmosphere of truth and right and God, and will be consciously inspired by these divine, omnipotent motives.

But who will compose this future race? We cannot tell. And yet the attempt to answer the question may open our eyes to truth of great practical importance.

It would seem to be a fact that the offspring of a cross between different races of the same species is as a rule more vigorous than that of either pure race. Human history seems to show the same result. The English race is a mixture of Celts, Anglo-Saxons, Danes, and Normans, with a sprinkling of other races. And a new fusion of a great number of most diverse strains is rapidly going on in the newly populated portions of America and in Australia. The mixture contains thus far almost purely occidental races. It will in

236 THE WHENCE AND THE WHITHER OF MAN

future almost certainly contain oriental also. For the races of India, Japan, and even China, are no farther from us to-day than the ancestors of many of our occidental fellow-citizens were a century ago. Racial prejudices, however strong, weaken rapidly through intercourse and better acquaintance. One of the grandest and least perceived results of missionary work is the preparation for this great fusion.

Many races will undoubtedly go down before the advance of civilization and have no share in the future. Progress seems to be limited to the inhabitants of temperate zones; and even here the weaker may be crowded out before the stronger rather than absorbed by them. But many whom we now despise may have a larger inheritance in the future than we. ·God is clearly showing us that we should not count any man, much less any nation, common or unclean. And the laws of evolution give us a firm confidence that no good attained by any race or civilization will fail to be preserved in the future.

The forms which seem to us at any one time the highest are as a rule not the ancestors of the race of the future. These highest forms are too much specialized, and thus fitted to a narrow range of space, time, and general conditions; when these change they pass away. Specialization is doubly dangerous when it follows a wrong line. But whenever it is carried far enough to lead to a one-sided development, it narrows the possibility of future advance; for it neglects or crowds out or prevents the development of other powers essential to life. The mollusk neglected nerve and muscle. But the scholar may, and often does, cultivate the brain at the expense of the rest of the body

until he and his descendants suffer, and the family becomes extinct.

The young men of the nobility of wealth, birth, and
fashion usually marry heiresses, if they can. But only
in families of enormous wealth can there be more than
one or two heiresses in the same generation. She
has very probably inherited a portion of her wealth
from one or more extinct branches of the family.
Moreover, not to speak of other factors, the labor and
anxiety which have been essential to the accumulation
and preservation of these great fortunes, or the mode
of life which has accompanied their use or abuse, tend
to diminish the number of children. Heiresses to very
large fortunes usually therefore belong to families
which are tending to sterility. And this has very probably been no unimportant factor in the extinction of
"noble " families.

A sound body contains many organs, all of which
must be sound. And in a sound mind there is an
even greater number of faculties, all of which must be
kept at a high grade of efficiency. Man is a marvellously complex being, and more in danger of a narrow
and one-sided development than any lower animal.
And it is very easy for a certain grade or class of society, or for a whole race, to become so specialized, by
the cultivation of only one set of faculties as to altogether prevent its giving birth to a complete humanity.
Along certain broad lines the Greeks and Romans attained results never since equalled. But their neglect
of other, even more important, powers and attainments, especially the moral and religious, doomed
them to a speedy decay. The rude northern races
were on the whole better and nobler, and became heirs

238 *THE WHENCE AND THE WHITHER OF MAN*

to Greek art and letters, and to Roman law. And this is another illustration of the advantage or necessity of the fusion of races.

To answer the question, " Which stratum or class in the community or world at large is heir to the future ? " we must seek the one which is still to a large extent generalized. It must be maintaining, in a sound body, a steady, even if slow, advance of all the mental powers. It will not be remarkable for the high development or lack of any quality or power ; it must have a fair amount of all of them well correlated. It must be well balanced, " good all around," as we say. And this class is evidently neither the highest nor the lowest in the community, but the " common people, whom God must have loved, because he made so many of them."

They have, as a rule, fair-sized or large families. Their bodies are kept sound and vigorous by manual labor. They are compelled to think on all sorts of questions and to solve them as best they can. They have a healthy balance of mental faculties, even if they are not very learned or artistic. They are kept temperate because they cannot afford many luxuries. Their healthy life prevents an undue craving for them. They help one another and cultivate unselfishness. The good old word, neighbor, means something to them. They have a sturdy morality, and you can always rely upon them in great moral crises. They are patriotic and public-spirited ; they have not so many, or so enslaving, selfish interests. They have always been trained to self-sacrifice and the endurance of hardship ; and heroism is natural to them. They have a strong will, cultivated by the battle of daily life. And among them religion never loses its hold.

But what of our tendencies to specialization in education and business ? Are these wrong and injurious ? Specialization, like ·great wealth, is a great danger and a fearful test of character. It tends to narrowness. If you will know everything about something, you must make a great effort to know something about, and have some interest in, everything. The great scholar is often anything but the large-minded, whole-souled man which he might have become. He has allowed himself to become absorbed in, and fettered by, his specialty until he can see and enjoy nothing outside of it. There is no selfishness like that of learning.

We can accomplish nothing unless we concentrate our efforts upon a comparatively narrow line of work. But this does not necessitate that our views should be narrow or our aims low. Teufelsdröckh may live on a narrow lane ; but his thoughts, starting along the narrow lane, lead him over the whole world. The narrowness of our horizon is due to our near-sightedness.

But the only absolutely safe specialization is the highest possible development of our moral and religious powers. For their cultivation only enlarges and strengthens all the other powers of body and mind. "But," you will object, "does religion always broaden ? " Yes. That which narrows is the base alloy of superstition. But a religion which finds its goal and end in conformity to environment, character, and godlikeness can only broaden.

But there is the so-called "breadth" of the shallow mind which attempts to find room at the same time for things which are mutually exclusive. God and Baal, right and wrong, honesty and lying, selfishness and love, these are mutually exclusive. You cannot find room in your mind for both members of the pair at the same time. You must choose. And, when you have chosen, abide by your choice. A ladleful of thin dough fallen on the floor is very broad. But its breadth is due to lack of consistency. Better narrowness than such breadth.

But while individual specialization may be safe for the individual, and beneficial to the race, the race which is to inherit the future must remain unspecialized. It must not sacrifice future possibilities to present rapidity of advance. And the common people are advancing safely, slowly, but surely. Wealth and learning become of permanent prospective and real value only when they are invested in the masses. They are the final depositaries of all wealth—material, intellectual, moral, and religious. Whatever, and only that which, becomes a part of their life becomes thereby endowed with immortality. Will we invest freely or will we wait to have that which we call our own wrested from us? If we refuse it to our own kin and nation, it will surely fall to foreigners. "God made great men to help little ones."

The city of God on earth is being slowly "builded by the hands of selfish men." But the builders are becoming continually more unselfish and righteous, and as they become better and purer its walls rise the more rapidly.

CHAPTER IX

WE have studied the teachings of science concerning man and his environment, let us turn now to the teachings of the Bible. And though eight chapters have been devoted to the teachings of science, and only one to the teachings of the Bible, it is not because I underestimate the importance of the latter. It is more difficult to clearly discover just what are the teachings of Nature in science. The lesson is written in a language foreign to most of us, and one requiring careful study; and yet once deciphered it is clear. Science attains the laws of Nature by the study of animal and human history. But this record is a history of continually closer conformity to environment on the part of all advancing forms. The animal kingdom is the clay which is turned, as Job says, to the seal of environment, and it makes little difference whether we study the seal or the impression; we shall read the same sentence. Environment has stamped its laws on the very structure of man's body and mind. And the old biblical writers read these laws, guided by God's Spirit, in their own hearts, and in those of their neighbors, and in their national history, as the record of God's working, and gave us concrete examples of the results of obedience and disobedience. Hence the teaching of the Bible is always clear and unmistakable.

16

The Bible treats of three subjects—Nature, Man, and God—and the relations of each of these to the others. I have tried to present to you in the first chapter the biblical conception of Nature and its relation to God. In its relation to man it is his manifestation to us, and, in its widest sense, the sum of the means and modes through which he develops, aids, and educates us. And in this conception I find science to be strictly in accord with scripture.

Now what is the scriptural idea of man? Man interests us especially in three aspects. He is a corporeal being; he is an intellectual being; he is a moral being, with feelings, will, and personality.

Man's body. Plato considered the body as a source of evil and a hindrance to all higher life. And Plato was by no means alone in this. The Bible takes a very different view. Neglect of the body is always rebuked. The only place, so far as I can find, where the body is called vile is where it is compared with the glorious body into which it is to be transformed. "Your bodies," writes Paul to the Corinthians, "are members of Christ," "temples of the Holy Ghost." But the Bible teaches that the body is to be the servant, not the ruler, of the spirit. "I keep under my body, and bring it into subjection," continues Paul. Here again science is strictly in accord with scripture.

Man is an intellectual being. I need not quote the praises of knowledge in the Old Testament. They must be fresh in your mind. But the practical Peter writes, "giving all diligence add to your faith virtue; and to virtue knowledge." And Paul prays that the love of the Ephesians may " abound more and more in knowledge and in all judgment." But the important knowledge is the knowledge of God, and of Jesus Christ, our Lord and Master. And similarly science emphasizes that the chief end of all knowledge is that we should know the environment to which we are to conform. Knowledge is useful to strengthen and clarify the mind, that it may see and conform to truth and God: and if it fails to become a means to conformity, it has failed of the chief, and practically the only, end for which it was intended. We are to come "in the unity of the faith and of the knowledge of the Son of God, unto a perfect man, unto the measure of the stature of the fulness of Christ." But knowledge which only puffs up and distracts the mind from the great aims and ends which it should serve is rebuked with equal emphasis by the Bible and by science.

I would not claim that we have set too high a value upon knowledge, perhaps we cannot; but there is something far higher on which we are inclined to set far too low a value. This is righteousness and love; and true wisdom is knowledge permeated, vivified, and transfigured by devotion to these higher ends. And in this highest realm of the mind feeling and will rule conjointly. Love is a feeling which always will and must find its way to activity through the will, and it is an activity of the will roused by the very deepest feeling, inspired by a worthy object. If you try to divorce them, both die. Hence Paul can say, "Though I speak with the tongues of men and of angels, and though I have the gift of prophecy, and understand all mysteries and all knowledge; and though I have all faith, so that I could remove mountains, and have not love, I am nothing." And John goes, if possible, even farther and says, "Every one that loveth is born of

God, and knoweth God. He that loveth not, knoweth not God; for God is love." And this sort of love bears and believes and hopes and endures, and never fails. And for this reason the Bible lays such tremendous emphasis on the heart, not as the centre of emotion alone, but as the seat of will as well. And science points to the same end, though she sees it afar off.

And what of God? God is a Spirit, Creator, Author, and Finisher of all things, and filling all. But while omnipotent, omnipresent, and omniscient, these are not the characteristics emphasized in the Bible. He is righteous. "Shall not the judge of all the earth do right?" is the grand question of the father of the faithful. And when Moses prays God to show him his glory, God answers, "I will make all my goodness pass before thee." He is the "refuge of Israel," the "everlasting arms" underneath them, pitying them "as a father pitieth his children." And in the New Testament we are bidden to pray to our Father, who *is* love, and whose temple is the heart of whosoever will receive him. Truly a very personal being.

Now the Bible rises here indefinitely above anything that mere natural science can describe. But can the ultimate "Power, not ourselves, which makes for righteousness" and unselfishness, of whose presence in environment science assures us, be ever better described than by these words concerning the "Father of our spirits?"

And an infinitely wise, good, and loving being will have fixed modes of working; for "with him is no variableness, neither shadow of turning." Thus only can man trust and know him. The old Stoic philosopher tells us "everything has two handles, and can be carried by one of them, but not by the other." So with God's laws. Many seem to look upon them as a hindrance and limitation to him in carrying out his righteous and loving will toward man. But they are really the modes or means of his working, which he uses with such regularity and consistency that we can always rely upon them and him. The pure river of the water of life proceedeth from the throne of God and of the Lamb.

If I am lying ill waiting anxiously for the physician I can think of this great city as a mass of blocks of houses separating him from me. But the houses have been arranged in blocks so as to leave free streets, along which he can travel the more quickly. And God's laws are not blocks, but thoroughfares, planned that the angels of his mercy may fly swiftly to our aid. We are prone to forget that these laws are expressly made for your and my benefit, as well as that of all beings, that we may be righteous and unselfish. And this is one ground of the apostle's faith that " all things work together for good to them that love God." And in the Apocalypse the earth helps the woman. It must be so.

But what if you or I try to block the thoroughfare? What would happen to us if we tried to stop bare-handed the current of a huge dynamo, or to hold back the torrent of Niagara? Nothing but death can result. And what if I stem myself against the " river of the water of life, proceeding from the throne of God," and try to turn it aside or hold it back from men perishing of thirst? And that is just what sin is, even if done carelessly or thoughtlessly ; for men have no right to be careless and thoughtless about some things. "The wages of sin is death;" physical death for breaking physical law, and spiritual death for breaking spiritual law. How can it be otherwise? The wages are fairly earned. The hardest doctrine for a scientific man to believe is that there can be any forgiveness of such sin as the heedless, ungrateful breaking of such wise and beneficent laws of a loving Father. And yet my earthly father has had to forgive me a host of times during my boyhood. Perhaps I can hope the same from God; I take his word for it.

But if you or I think that it is safe to trifle with God's laws, we are terribly mistaken. The Lord proclaimed himself to Moses as "The Lord, the Lord God, merciful and gracious, long-suffering, and abundant in goodness and truth, keeping mercy for thousands, forgiving iniquity and transgression and sin, and that will by no means clear the guilty; visiting the iniquity of the fathers upon the children, and upon the children's children, unto the third and to the fourth generation." But someone will say, This is terrible. It is terrible; but the question is, Does the Bible speak the truth about nature? Is nature a "fairy godmother," or does she bring men up with sternness and inflict suffering upon the innocent children, if necessary, lest they copy after their sinful parents? Do the children of the defaulter and drunkard and debauchee suffer because of the sins of their father, or do they not? If the blessings won by parental virtue go down to the thousandth generation, must not the evil consequences of sin go down to the third or fourth?

That we are not under the law, but under grace,

does not mean, as some seem to think, that it is safe to sin. Otherwise the forgiveness of God becomes the lowest form of indulgence slanderously attributed to the Church of Rome. We gain freedom from law as well as penalty only by obedience. The artist can safely forget the laws and rules of his art only when by long obedience and practice he obeys them unconsciously. We seem to be threatened with a belief that God will never punish sin in one who has professed Christianity. This view cheapens sin and makes pardon worthless, it takes the iron out of the blood, and the backbone out of all our religion and ethics. It ruins Christians and disgraces Christianity. We sometimes seem to think that our nation or church or denomination is so important to the carrying on of God's work that he cannot afford to let any evil befall us, whatever we may do or be.

"Hear this, I pray you, ye heads of the house of Jacob, and princes of the house of Israel, that abhor judgment and pervert all equity. They build up Zion with blood, and Jerusalem with iniquity. The heads thereof judge for reward, and the priests thereof teach for hire, and the prophets thereof divine for money : yet will they lean upon the Lord and say, Is not the Lord among us? none evil can come upon us. Therefore shall Zion for your sake be ploughed as a field, and Jerusalem shall become heaps, and the mountain of the house as the high places of the forest." That was plain preaching, and the people did not like it. They would not like it any better to-day; it would come too near the truth.

But others seem to think that God is too kind, not to say good-natured, to allow his children to suffer for their sins.   This is part of a creed, unconsciously very widely held to-day, that comfort, not character, is the chief end of life.   Now if God is too kind to allow his children to suffer some of the natural consequences of sin, he is not a really kind and loving father, he is spoiling his children.   Salvation is soundness, sanity, health; just as holiness is wholeness, escape from the disease, and not merely from the consequences of sin.   A physician, unless a quack, never promises relief from a deep-seated disease without any pain or discomfort.   And if the disease is the result of indulgence, he warns us that relapse into indulgence will bring a worse recurrence of the pain. Perhaps, after all, Socrates was not so far from right when he maintained that if a man had sinned the best and only thing for him is to suffer for it.   " God the Lord will speak peace unto his people, and to his saints: but let them not turn again to folly." And our Lord says, " Think not that I am come to destroy the law or the prophets ; I am not come to destroy, but to fulfil.   For verily I say unto you, Till heaven and earth pass, one jot or one tittle shall in no wise pass from the law till all be fulfilled.   For I say unto you, That except your righteousness shall exceed the righteousness of the scribes and Pharisees, ye shall in no case enter into the kingdom of heaven." If we would be great in the kingdom of heaven we must do and teach the commandments.   One of the best lessons that the clergy can learn from science is that law and penalty are not things of the past.   They are eternal facts ; and if so, ought sometimes to be at least mentioned from the pulpit as well as remembered in the pew.

But if God is a person striving to communicate with man, and if man is a person intended to conform to environment by becoming like God, what is more probable from the scientific stand-point than that God should seek and find some means of making himself clearly known to man in some personal way? I do not see how any scientific man who believes in a personal God can avoid asking this question. And is there any more natural solution of the question than that given in the Bible? " God was in Christ reconciling the world to himself." "God, who spake in time past unto the fathers by the prophets, hath in these last days spoken unto us by his son." Philip says, " Lord, show us the Father and it sufficeth us." Jesus saith unto him, " Have I been so long time with you, and dost thou not know me, Philip? he that hath seen me hath seen the Father; how sayest thou shew us the Father? Believest thou not that I am in the Father and the Father in me? the words that I say unto you I speak not from myself: but the Father abiding in me doeth his works."

" And this is the condemnation, that light is come into the world, and men loved darkness rather than light, because their deeds were evil."

Something more is needed than light. We need more light and knowledge of our duty; we need vastly more the will-power to do it. I know how I ought to live ; I do not live thus. What I need is not a teacher, but power to become a son of God. " I delight in the law of God after the inward man : but I see a different law in my members, warring against the law of my mind, and bringing me into captivity under the law of sin which is in my members. O wretched man that I am! who shall deliver me out of the body of this death?"

This is the terrible question. How is it to be answered? Let us remember our illustration of the change wrought in that panic-stricken army before Winchester by the appearance of Sheridan. What these men needed was not information. No plan of battle reported as sure of success by trustworthy and competent witnesses, and forwarded from the greatest leader could have stayed that rout. What they needed was Sheridan and the magnetic power of his personality. This is the strange power of all great leaders of men, whether orators, statesmen, or generals. It is intellect acting on and through intellect, but it is also vastly more; it is will acting on will. The leader does not merely instruct others, he inspires them, puts himself into them, and makes them heroes like himself.

Now something like this, but vastly grander and deeper, seems to me to have been the work of our Lord. Read John's gospel and see how it is interpenetrated with the idea of the new life to be gained by contact with our Lord, and how this forms the foundation of his hope and claim to give men this new life by drawing them to himself. And Peter says that it was impossible for the Prince of Life to be holden of death, for he was the centre and source from which not only new thoughts and purposes, but new will and life was to stream out into the souls of men. This power of our Lord may have been miraculous and supernatural in degree; I feel assured that it was not unnatural in kind and mode of action.

And here, young men, pardon a personal word

about your preaching. You will need to preach many sermons of warning against, and denunciation of, sin; many of instruction in duty. The Bible is a storehouse of instruction and men need it, and you must make it clear to them. All this is good and necessary, but it is not enough. Learn from the experience of the greatest preacher, perhaps, who ever lived.

Paul, the greatest philosopher of ancient times, came to Athens. You can well imagine how he had waited and longed for the opportunity to speak in this home of philosophy and intellectual life. Now he was to speak, not to uncultured barbarians, but to men who could understand and appreciate his best thoughts. He preached in Athens the grandest sermon, as far as argument is concerned, ever uttered. I doubt if ever a sermon of Paul's accomplished less. He could not even rouse a healthy opposition. The idea of a new god, Jesus, and a new goddess, the Resurrection, rather tickled the Athenian fancy. He left them, and, in deep dejection, went down to Corinth. There he determined to know only· "Christ and him crucified," and thus preaching in material, vicious Corinth he founded a church.

Some of you will go through the same experience. You will preach to cultured and intelligent audiences, and they will listen courteously and eagerly as long as you tell them something new, and do not ask them to do anything. The only possible way of reaching Athenian intellect or Corinthian materialism and vice is by preaching Christ, "the power of God and the wisdom of God." And you will reach more Corinthians than Athenians."

You may preach sermons full of the grandest philosophy and theology, and of the highest, most exact, science; you may chain men by your logic, thrill them by your rhetoric, and move them to tears by your eloquence, and they will go home as dead and cold as they came. What they need is power, life. But preach " Christ and him crucified "—not merely dead two thousand years ago—but risen and alive for evermore, and with us to the end of the world, the grandest, most heroic, divinest helper who ever stood by a man, one all-powerful to help and who never forsakes, and every one of your hearers who is not dead to truth will catch the life, and go home alive and not alone.

So long as we preach a dead Christ we shall have a dead church, as hopeless as the apostles were before the resurrection. " But now is Christ risen from the dead," " alive for evermore." See how Paul and Peter and John, and doubtless all the others, talked with him and he with them, after he was taken from them, and you have found the secret of their power, and of that of all the great Christian heroes and martyrs who could truly say, Lord Jesus, we understand each other. Better yet, prove by experience that it is possible for every one of us.

And our Lord and Master is the connecting link between God and man, through whom God's own Holy Spirit is poured like a mighty flood into the hearts and lives of men, transfiguring them and filling them with the divine power. This is the biblical idea of Christianity; man, through Christ, flooded and permeated and interpenetrated with the Holy Spirit of God. And thus Paul is dead and yet alive, but fully possessed and dominated by the spirit of Christ. Alive as never before, and yet his every thought, word, and deed is really that of his great leader. Can you talk of self-denial to such a Christian? He had forgotten that such a man as Saul of Tarsus or Paul ever existed; he lives only in his Master's work, and is transfigured by it. This, and nothing less, is Christianity, and this is the very highest and grandest heroism. Paul conquers Europe single-handed, alone he stands before Cæsar's tribunal, and yet he is never alone; and from the gloom of the Mammertine dungeon he sends back a shout of triumph. And Peter walks steadily, cheerfully, and unflinchingly, in the footsteps of his Master to share his cross.

Let us, before leaving this topic, notice carefully just what religion, and especially Christianity, is not.

1. It is not merely opinion or intellectual belief in a creed. This may be good, or even necessary, but it is not religion. "Thou believest that there is one God; thou doest well: the devils also believe and tremble." We speak with pride, sometimes, of our puissant Christendom, so industrious, so intelligent, so moral, with its ubiquitous commerce, its adorning arts, its halls of learning, its happy firesides, and its noble charities. And yet what is our vaunted Christendom but a vast assemblage of believing but disobedient men? Said William Law to John Wesley, "The head can as easily amuse itself with a living and justifying faith in the blood of Jesus as with any other notion." The most sacred duty may degenerate into a dogma, asking only to be believed. "I go, sir," answered the son in the parable, "but went not."

2. It is not mere feeling. It is neither hope of heaven's joy, nor fear of hell's misery. It may rightly

include these, but it is vastly more and higher. It is neither ecstasy nor remorse. The most resolutely impenitent sinner can shout " Hallelujah," and " Woe is me," as loudly as any saint. Now feeling is of vast importance. It stands close to the will and stimulates it, but it is not conformity. The will must be aroused to a robust life.

3. Christianity is these and a great deal more. Mere belief would make religion a mere theology. Mere emotion would make it mere excitement. The true divine idea of it is a life; doing his will, not indolently sighing to do it, and then lamenting that we do it not; but the thing itself in actual achievement, from day to day, from month to month, from year to year. Thus religion rises on us in its own imperial majesty. It is no mere delight of the understanding in the doctrines of our faith ; no mere excitement of the sensibilities, now harrowed by fear, and now jubilant in hope ; but a warfare and a work, a warfare against sin, and a work with God. Religion is not an entertainment, but a service. We are to set before us the perfect standard, and then struggle to shape our lives to it. Personal sanctity must be made a business of.*

A little more than thirty years ago a regiment was sent home from the Army of the Potomac to enforce the draft after the riots in this city. Some of you may picture to yourselves a thousand men with silk banners and gold lace and bright uniforms, resplendent in the sunshine. You could not make a worse mistake.

First in that gray early morning came two old flags, so torn by shot and shell that there was hardly enough

---

* This page is mainly a series of quotations from Dr. R. D. Hitchcock's sermon on " Religion, the Doing of God's Will."

left of them to tell whether the State flag was that of Massachusetts or Virginia. And behind these came scant three hundred men. All the rest were sleeping between Washington and Richmond, some on almost every battle-field. The uniforms were old and faded from sun and rain. Only gun-barrel and bayonet were bright. And the men were scarred and tired and foot-sore, haggard from hard fighting and long, swift marches. For these men had been trained to be hurried back and forth behind the long line of battle, that they might be hurled into it wherever the need was greatest. I do not suppose that one of them could have delivered a fourth-of-July oration on Patriotism. They were trained not to talk, but to obey orders. But they had stood in the "bloody angle" at Spottsylvania all day and all night; and in the gray dawn of the next morning, when strength and courage are always at ebb, faint and exhausted, their last cartridge shot away, had sprung forward at the command of their colonel to make a last desperate, forlorn defence with the bayonet against the advancing enemy. Numbers do not count against men like these. What made them such invincible heroes? It was mainly the resolute will and long training to obey orders. A Christian should never forget that he is a soldier in the army of the Lord of Hosts; that enlistment is easy and quickly accomplished; but that the training is long, and that he must learn, above all, to "endure hardness."

And so, my brothers, I beg of you to preach a heroic Christianity, for if there ever was a heroic religion it is ours. If you offer merely free transportation to a future heaven of delight on "flowery beds of ease," you will enlist only the coward and the sluggard. But everyone who has a drop of strong old Norse blood in his veins will prefer a heathen Valhalla, though builded in hell, to such a heaven. And his Norse instincts will be nearer truth than your counterfeit of a debased Christianity. But preach the city of God's righteousness on earth and now among men, and call on every heroic soul to take sides with God against sin within himself and the evil and misery all around him. There is an almost infinite amount of strength, endurance, and heroism in this "slow-witted but long-winded" human race waiting to leap up at the appeal to fight once more and win a victory after repeated defeats before the sun goes down. Appeal to this and point to the great "captain of our salvation made perfect through sufferings," and every man that is of the truth will hear in your voice the call of the Master and King. You will not be disappointed, but among the publicans and fishermen of America you will find heroic souls, who will leave all to follow, as faithfully and unflinchingly as those from the shores of Galilee.

And what of faith? Faith is the personal attachment of a soul to such a leader. Fortunately the Bible contains a scientific monograph on this subject. I refer, of course, to the eleventh chapter of the epistle to the Hebrews. And the whole result is summed up in a few words of the thirteenth verse. The great heroes, like Enoch, Noah, and Abraham, "saw the promises afar off, and were persuaded of them, and embraced them, and confessed that they were strangers and pilgrims on the earth."

They saw the promises afar off, dimly, on the horizon of their mental vision; as one looks into the distance and cannot tell whether what he sees be cloud or mountain. And until they could make up their minds that there was some substance in the vision, they did not embrace it. They were not credulous. Neither were they carelessly or heedlessly sure that there was and could be nothing in the vision but mist and fancy. They recognized that on their decision of the question hung the life of which they meant to make the very most. They looked again and again, and kept thinking about it. Thus they became and were "persuaded of them." And most people stop here with a merely intellectual faith in their heads, and very little in their hearts and lives. Not so these old heroes ; they were not so purely and coldly intellectual that they could not *do* anything. They "embraced them." They said, that is exactly what I want and need, and I'll have it, if it costs me my life.

Now a promise is always conditional ; if you want one thing, you must give up something else. It involves a choice between alternatives ; you can have either one freely, you cannot have both. It was to them as to Christ on the "exceeding high mountain," God or the world ; God with the cross, or the world with Satan thrown in. And the same alternative confronts us.

Moses could be a good Jew or a good Egyptian. Most of us, while resolved to be excellent Jews at heart, would have said nothing about it, but remained sons of Pharaoh's daughter in order to benefit the Jews by our influence in our lofty station. We should have become miserable hybrids with all the vices and weaknesses of both races, but with none of the virtues of either. And for all that we should ever have done

17

the Jews might have rotted in Egyptian bondage. En-largement and deliverance would have arisen to the Jews from some other place; but we and our father's house would have been destroyed. By faith Moses refused to be called the son of Pharaoh's daughter, choosing rather to suffer affliction with the children of God, etc. And certainly he did suffer for it.

They embraced the promises with their whole hearts. They were stoned and sawn asunder rather than give them up. And what was the effect on their characters? Having counted the cost, and being perfectly willing to accept any loss or pain for the sake of these prom-ises, and hence inspired by them, they became sub-lime heroes. Through faith they " subdued kingdoms, wrought righteousness, obtained promises, stopped the mouths of lions, quenched the violence of fire, escaped the edge of the sword, out of weakness were made strong, waxed valiant in fight, turned to flight the armies of the aliens. And others had trials of cruel mockings and scourgings, yea, moreover of bonds and imprisonment : they wandered about in sheepskins and in goatskins ; being destitute, afflicted, tormented. Of whom the world was not worthy." That is a faith worth having, and it is as sound philosophy as it is scripture.

"These all died in faith, not having received the promises." Did they receive nothing? Moses and Elijah, Gideon and Barak gained power and heroism greater than we can conceive of. Surely that was enough. But they did not get the whole of the prom-ise, or even the best of it. And the simple reason was that God cannot make a promise small enough to be completely fulfilled to a man in his earthly life.

He gets enough to make him a king, but this does not
begin to exhaust the promise. It is inexhaustible.
This is the experience of anyone who will faithfully
try it. And this experience is the grandest argument
for immortality.

Therefore, "giving all diligence, add to your faith
virtue (ἀρετή, strength), and to virtue knowledge, and
to knowledge temperance (ἐγκράτεια, self-control), and
to temperance patience (ὑπομενῆ, endurance), and to
patience godliness, and to godliness brotherly kindness, and to brotherly kindness charity " (love).

And what of prayer? How can it be answered in
a universe of law? We certainly could have no confidence that our prayers could or would be answered
if ours were not a universe of law. God's laws are, as
we have seen, his modes of working out his great plan.
And the last and highest unfolding of God's plan is
the development of man. And man is to become conformed to his environment, and conformity of man's
highest powers to his environment is likeness to God.

The laws of nature, then, are in ultimate analysis
and highest aim the different steps in God's plan of
man's salvation from the disease of sin, not merely or
mainly from its consequences, and his attainment of
holiness. For this is the only true and sound manhood. Salvation is spiritual health, resulting also in
health of body and of mind. If God's laws are his
modes of carrying out his plan for godlikeness in man,
then they are so thought out as to be the means of
helping me to every real good.

The Bible declares explicitly that the aim of prayer
is not to inform God of our needs. For he knows
them already. It is not to change God's purpose,

for he is unchangeable, and we should rejoice in this. We are to pray for our daily bread; we are to pray for the sick; and, if best for them and consistent with God's plan, they shall recover. Elijah prayed for drought and prayed for rain, and was answered. And Abraham's prayer would have saved Sodom, had there been ten righteous men in the city. " Men ought alway to pray and not to faint."

> " More things are wrought by prayer
> Than this world dreams of. Wherefore let thy voice
> Rise like a fountain for me night and day.
> For what are men better than sheep or goats
> That nourish a blind life within the brain,
> If, knowing God, they lift not hands of prayer
> Both for themselves and those who call them friend?
> For so the whole round earth is every way
> Bound by gold chains about the feet of God."

But could not all these things be brought about without a single prayer? Not according to the plan of man's education which God has adopted. Whether he could well have made a plan by which material blessings could have been bestowed upon men who do not ask for them, I do not know. The ravens and all animals are fed without a single prayer, for they are not fitted or intended to hold communion with God. But a prayerless race of men has never been fed long; it has soon ceased to exist. God's plan of salvation and ordering of the universe involves prayer as a means of blessing and good things as an answer to prayer. God says, I make you a co-worker with me. I will help you in everything; but you must call on me for help, or you will forget that I am the source of your help and strength, and thus having lost your communion with me will die. " When Jeshurun waxed fat he kicked." This is the oft-repeated story of the Old Testament and of all history. And thus, while material blessings are given in answer to prayer, these are not the chief end for which prayer is to be offered.

Prayer is a means of conformity to environment, of godlikeness. How do you become like a friend? Of course by associating and talking with him. And why does it help you to associate with a hero? Simply because you cannot be with him without being inspired with his heroism. And so while I may pray for bread and clothes and opportunities, and God will give me these or something better; I will, if wise, pray for purity, courage, moral power, heroism, and holiness. And I know that these will stream from his soul into mine like a great river. And so I may pray for bread and be denied; for hunger, with some higher good, may be far better for me than a full stomach. But if I pray for any spiritual gift, which will make me godlike, and on which as an heir of God I have a rightful claim, every law and force in God's universe is a means to answer that prayer. And best of all, if I pray for the gift of God's Spirit, that is the prayer which the whole world of environment has been framed to answer.

But this I can never have unless I hunger for it. I can never have it to use as a means of gaining some lower good which I worship more than God. God will not and cannot lend himself to any such idolatry. I must be willing to give up anything and everything else for its attainment. Otherwise the answer to the prayer would ruin me.

I cannot grasp the higher while using both hands to grasp the lower.

Thus religion is the interpenetration and permeation of my personality by that of God. And prayer is the communion by which this permeation becomes possible. And faith is the vision of these possibilities, the being persuaded by them, and the resolute purpose to attain them. And faith in Christ is confiding communion with him and obedience to his commands that his divine life may flow over into me and dominate mine. And common-sense, and the more refined common-sense which we call science, can show me no other means to the attainment of that godlikeness which is the only true conformity to environment.

And, holding such a belief and faith, we must be hopeful. And only next in importance to faith and love stands hope. The hero must be hopeful. And when times look dark about you, and they sometimes will, you must still hope.

> "O it is hard to work for God,
>   To rise and take his part
> Upon the battle-field of earth,
>   And not sometimes lose heart!
>
> "O there is less to try our faith
>   In our mysterious creed,
> Than in the godless look of earth
>   In these our hours of need.
>
> "Ill masters good; good seems to change
>   To ill with greatest ease;
> And, worst of all, the good with good
>   Is at cross purposes.

" Workman of God! O lose not heart,
  But learn what God is like ;
  And in the darkest battle-field
  Thou shalt know where to strike.

" Muse on his justice, downcast soul !
  Muse, and take better heart ;
  Back with thine angel to the field,
  Good luck shall crown thy part !

" For right is right, since God is God ;
  And right the day must win ;
  To doubt would be disloyalty,
  To falter would be sin."

Hope on, be strong and of a good courage. For in the dark hours others will lean on you to catch your hope and courage. To many a poor discouraged soul you must be " a hiding-place from the wind and a covert from the tempest ; as rivers of water in a dry place, as the shadow of a great rock in a weary land." Every power and force in the universe of environment makes for the ultimate triumph of truth and right. Defeat is impossible. "One man with God on his side is the majority that carries the day. 'We are but two,' said Abu Bakr to Mohammed as they were flying hunted from Mecca to Medina. 'Nay ;' answered Mohammed, 'we are three; God is with us.'"

And not only the race will triumph and regain the Paradise lost. The city of God shall surely be with men, and God will dwell with them and in them. But you and I can and shall triumph too.

We are prone to feel that the individual man is too insignificant a being to be the object of God's care and forethought. But we should not forget that it is the individual who conforms, and that the higher and nobler race is to be attained through the elevation of individuals, one after another. God deals with races and nations as such. But his laws and promises are made almost entirely for the individuals of which these larger units are concerned.

But there is another standpoint from which we may gain a helpful view of the matter. I may be the meanest citizen of my native state, and my father may leave me heir of only a few acres of rocky land. But, if my title is good, every power in the state is pledged to put me in possession of my inheritance. They who would rob me may be strong; but the state will call out every able-bodied man, and pour out every dollar in its treasury before it will allow me to be defrauded of my legal rights. And it must do this for me, its meanest citizen, else there is no government, but anarchy, and oppression, and the rule of the strongest. And we all recognize that this is but right and neces- sary, and would be ashamed of our state and govern- ment were it not literally true.

If I travel in distant lands, my passport is the sign that all the power of these United States is pledged to protect me from injustice. Think of the sensitiveness of governments to any wrong done to their private citizens. England went to war with Abyssinia to pro- tect and deliver two Englishmen. And shall God do less? Can he do less? If it is only just and right and necessary for earthly governments to thus care for their citizens, shall not the ruler and " judge of all the earth do right?"

Now you and I are commanded to be heirs of God, to attain to likeness to him. This is therefore our

legal right, guaranteed by him, for every command of
God is really a promise. And he will exhaust every
power in the universe before he allows anything to
prevent us from gaining our legal rights, provided
only that we are earnest in claiming them.

But if I alienate my rights to my inheritance, the
commonwealth cannot help me. If I renounce my
citizenship, the government of the United States can
no longer protect me. And so I can alienate my
" right to the tree of life," and to entrance into the
city, and I can forfeit my heirship to all that God
would give me. " For I am persuaded that neither
death, nor life, nor angels, nor principalities, nor things
present, nor things to come, nor powers, nor height,
nor depth, nor any other creation, shall be able to
separate us from the love of God, which is in Christ
Jesus our Lord." But I can alienate and make void
every promise and title, if I will or if I do not care.
This is the unique glory and awfulness of the human
will. And we know that to them that love God all
things work together for good. " If God is for us
who is against us? " It must be so if God's laws
are his modes of aiding men to conform to environment.

And what of the church? Is it anything else or
other than a means of aiding man to conform to environment? If it fails of this, can it be any longer the
church of God? The church is a means, not an end.
And it is a means of godlikeness in man.

Some would make it a social club. The bond of
union between its members is their common grade of
wealth, social position, or intellectual attainments.
And this idea of the church has deeper root in the

minds of us all than we think. I can imagine a far better club than one formed and framed on this principle, but it is difficult for me to imagine a worse counterfeit of a church. Others make it a source of intellectual delectation, and the means of hearing one or two striking sermons each week. Such a church will conduce to the intelligence of its members, and may be rather more, though probably less, useful than the old New England Lyceum lecture system. Such a church is of about as much practical value to the world at large as some consultations of physicians are to their patients. The doctors have a most interesting discussion, but the patient dies, and the nature of the disease is discovered at the autopsy. Others still would make of the church a great railroad system, over which sleeping-cars run from the City of Destruction, with a coupon good to admit one to the Golden City at the other end. The coaches are luxurious and the road-bed smooth. The Slough of Despond has been filled, the Valley of Humiliation bridged at its narrowest point, and the Delectable Mountains tunnelled. But scoffers say that most of the passengers make full use of the unlimited stop-over privileges allowed at Vanity Fair.

The Bible would seem to give the impression that the church is the army of the Lord of Hosts, a disciplined army of hardy, heroic souls, each soldier aiding his fellow in working out the salvation which God is working in him. And it joins battle fiercely and fearlessly with every form of sin and misery, counting not the odds against it. And the Salvation Army seems to me to have conceived and realized to a great extent just what at least one corps in this grand army can and should be. And you and I can learn many a lesson from them.

The church is the body of which Christ is the head, and you and I are " members in particular." Let us see to it that we are not the weak spot in the body, crippling and maiming the whole. The church is the city of God among men, and we are its citizens, bound by its laws, loyal servants of the Great King, sworn to obey his commands and enlarge his kingdom, and repel all the assaults of his adversaries. Thus the Bible seems to me to depict the church of God. But what if the army contains a multitude of men who will not obey orders or submit to discipline? or if the city be overwhelmed with a mass of aliens, who see in its laws and institutions mainly means of selfish individual advantage? Responsibility, not privilege, is the foundation of strong character in both men and institutions. There was a good grain of truth in the old Scotch minister's remark, that they had had a blessed work of grace in his church; they had not taken anybody in, but a lot had gone out.

There are plenty of churches of Laodicea to-day. May you be delivered from them. But, thank God, there are also churches of Philadelphia and Smyrna. May you be pastors of one of the latter. It will not pay you a very large salary, for Demas has gone to the church of Laodicea, because the minister of the church of Smyrna was not orthodox, or not sufficiently spiritually minded—meaning thereby that he rebuked the sins of actual living men in general, and of Demas in particular—or preached politics, and did not mind his business. And your church may be small. For many of the congregation have gone to the church around the other corner, which is mainly a cluster of associations, having excellent names, and useful for almost every purpose except building up a manly, rugged, heroic, godlike character. The minister there, they will tell you, preaches delightful sermons. They make you "feel so good." He annihilates pantheism, and his denunciations of materialism are eloquent in the extreme. But his incarnations of materialism are Huxley and Darwin, and to the uncharitable he seems to almost carefully avoid any language which might seem to reflect upon the dollar- and place-worship of some of the occupants of his front pews. Now, I am not here to defend Mr. Huxley or Mr. Darwin. Withstand them to the face wherever they are to be blamed. And for some utterances they are undoubtedly to be blamed, honest souls as they were. But I for one cannot help feeling that there is among the "dwellers in Jerusalem" a materialism of the heart which is indefinitely worse than any intellectual heresy. When you hit at the one heresy strike hard at the other also.

Many will have left your little church of Smyrna. It had to be so. For the divine sifting process, which is natural selection on its highest plane, has not ceased to work. It must and shall still go on; it cannot be otherwise. Has the great principle ceased to be true in modern history that "though the number of the children of Israel be as the sand of the sea, a remnant shall be saved?"

But do not be discouraged. Preach Christ and a heroic Christianity. Do not be afraid to demand great things of your people. Remember that Ananias was encouraged to go to Paul because the Lord would show Paul how great things he should suffer for the name of Jesus. This is what appeals to the heroic in every man, and we do not make nearly enough use of it. And the heroic Christ and his heroic Christianity will draw every heroic soul in the community to himself. They may not be very heroic looking. You may be in some hill town in old Massachusetts "Nurse of heroes." Pardon me, I do not intend to be invidious. Heroism is cosmopolitan. One of the pillars of your church may be the school-teacher of the little red school-house at the fork of the roads, in the yard ornamented with alders, mulleins, and sumachs. She boards around, and is clad in anything but silks and sealskins. But she trains well her band of hardy little fellows, who will later fear the multitude as little as they now mind the Berkshire winds. And from the pittance she receives for training these rebellious urchins into heroic men she is supporting an old mother somewhere, or helping a brother to an education. And your deacon will be some farmer, perhaps uncouth in appearance and rough of dress, and certainly blunt in his scanty speech. He'll not flatter you nor your sermons; and until you've lived with him for years you will not know what a great heart there is in that rugged frame, and what wealth of affection in that silent hand-shake. And there is his wife. She is round and ample, and certainly does not look especially solemn or pious. She is aunt and mother to the whole community, the joy of all the children, nurse of the sick, and comfort of the dying. She is doing the work of ten at home, and of a host in the village.

And your right-hand man is great Onesiphorus from the mill down in the valley, fighting an uphill battle to keep the wolf from the door, while he and his wife deny themselves everything, that their flock of children may have better training for fighting God's battles than they ever enjoyed.

I cannot describe these men and women. If you have lived with them, you will need no description, and would resent the inadequacy of mine. If you have never had the good fortune to live with them, it is impossible to make you see them as they are. When you once have thoroughly known them, language will fail you to do them justice, and you will prefer to be silent rather than slander them by inadequate portrayal. They are at first sight not attractive-looking. If you stand outside and look at them from a distance their lives will appear to you very humdrum and prosaic. But remember that for almost thirty years our Lord lived just such a life in Nazareth, making ploughs and yokes ; and then, when the younger brothers and sisters were able to care for themselves, snatched three years from supporting a peasant family in Galilee to redeem a world. And who was Peter but a rough, hardy fisherman ?

Now a Paul, trained at the feet of Gamaliel, was also needed ; and the twelve did not come from the lowest ranks of society. But they were honest, industrious, practical, courageous, hardy, common people. And single-handed they went out to conquer empires. And they succeeded through the power of God in them.

Who knows the possibilities of your little church in the hilltown of Smyrna ? These men and women are the pickets of God's great host. They are scattered up and down our land, fighting alone the great battle, unknown of men and sometimes thinking that they must be forgotten of God. And the picket's lonely post is what tries a man's courage and strength.

Take your example from Paul's epistle. Greet Phebe, the schoolmistress, and Aquila and Priscilla on their rocky farm on the mountain-side, and greet the burden-bearing Onesiphorus. And give them God's greeting and encouragement, for he sends it to them through you. Show them the heroism which there is in their "humdrum" lives; and cheer them in the efforts, of whose grandeur they are all unconscious. Bid them "be strong and of a very good courage." For in the character of these people there is the granite of the eternal hills, and in their hearts should be the sunshine of God. Do not be ashamed of your congregation. Their dimes or dollars may look pitifully small and few on the collector's plate; only God sees the real immensity of the gift in the self-denial which it has cost. Your people will take sides with the cause of right, while it is still unpopular. They have furnished the moral backbone and unswerving integrity of many of your great business houses in this city to-day. From those families will go forth the men whom the good will trust and the evil fear. The power for good proceeding from your church will be like the floods which Ezekiel saw pouring out from beneath the threshold of the Lord's house.

For these common people, whom "God must have loved because he made so many of them," are the true heirs to the future. And wealth and culture, art

and learning, are to burn like torches to light their
march. Finally, my young brothers, do not be bitterly disappointed if you are not "popular preachers."
Do not let too many people go to sleep under your
preaching, even if one young man did go to sleep under
one of Paul's sermons. But if now and then someone
is angry at what you have said, do not worry too
much over it. Preach the truth in love. If Elijah
and John the Baptist, and Peter and Paul, were to
preach to-day I doubt greatly whether they would be
popular preachers. I cannot find that they ever were
so. They would probably be peripatetic candidates,
until someone supported them as independent evan-
gelists. After their death we would rear them great
monuments, and then devote ourselves to railing at
Timothy because he was not more like what we imagine Paul was.

Even Socrates found that he must bid farewell to
what men count honors, if he would follow after truth.
You may have the same experience. You will have to
champion many an unpopular cause, and your people
will not like it. They will say you lack tact. Now
Paul was a man of infinite tact. Witness his sermon
on Mars' Hill. But if his letters to the church in Corinth were addressed to most modern churches, they
would soon set out in search of a pastor of greater
adaptability.

If you play the man, and fight the good fight of
faith, I do not see how you can always avoid hitting
somebody on the other side. And he will pull you
down if he can; and will probably succeed in sometimes making your life very uncomfortable. Remember the teaching of scripture and science, that the up-

ward path was never intended to be easy. The scriptural passages to this effect you can find all through the gospels and epistles, and I need not quote them to you. I will, however, .tell you honestly that many are of the opinion that these passages are now obsolete, being applicable only to the first centuries, or to especially critical times in the history of the church. I cannot share that view, but, lest I seem too old-fashioned, will merely quote the ringing words of our own Dr. Hitchcock, that "no man ever enters heaven save on his shield." And allow me to quote in the same connection the testimony of that prince of scientists, Professor Huxley, in his lecture on "Evolution and Ethics : "

" If we may permit ourselves a larger hope of abatement of the essential evil of the world than was possible to those who, in the infancy of exact knowledge, faced the problem of existence more than a score of centuries ago, I deem it an essential condition of the realization of that hope that we should cast aside the notion that the escape from pain and sorrow is the proper object of life.

" We have long since emerged from the heroic childhood of our race, when good and evil could be met with the same 'frolic welcome ;' the attempts to escape from evil, whether Indian or Greek, have ended in flight from the battle-field ; it remains to us to throw aside the youthful over-confidence and the no less youthful discouragement of nonage. We are grown men, and must play the man

. . . " 'strong in will
To strive, to seek, to find, and not to yield,'

18

cherishing the good that falls in our way and bearing the evil in and around us, with stout heart set on diminishing it. So far we all may strive in one faith toward one hope :

> " ' It may be that the gulfs will wash us down,
> It may be we shall touch the Happy Isles.

>   . . .   " but something ere the end,
> Some work of noble note may yet be done.' "

We must be strong and of a very good courage. While the avoidance of pain and discomfort, or even happiness, cannot be the proper end of life, it is not a world of misery or an essentially and hopelessly evil world. There is plenty of misery in the world, and we cannot deny it. Neither can we deny that God has put us in the world to relieve misery, and that until we have made every effort and strained every nerve as we have never yet done, we, and not God, are largely responsible for it. But behind misery stand selfishness and sin as its cause. And here we must not parley but fight. And the hosts of evil are organized and mighty. " The sons of this world are for their own generation wiser than the sons of light." And we shall never overcome them by adopting their means. But we can and shall surely overcome. For he that is with us is more than they that be with them. " The skirmishes are frequently disastrous to us, but the great battles all go one way." And we long for the glory of "him that overcometh." But the victor's song can come only after the battle, and be sung only by those who have overcome. And we would not have it otherwise if we could. The closing words of Dr. Hitchcock's last sermon are the following:

"It is one of the revelations of scripture that we are to judge the angels, sitting above them on the shining heights. It may well be so. Those angels are the imperial guard, doing easy duty at home. We are the tenth legion, marching in from the swamps and forests of the far-off frontier, scarred and battered, but victorious over death and sin."

CHAPTER X

IN all our study we have taken for granted the truth of the theory of evolution. If you are not already persuaded of this by the writings of Darwin, Wallace, and many others, no words or arguments of mine would convince you. We have used as the foundation of our argument only the fundamental propositions of Mr. Darwin's theory.

But while all evolutionists accept these propositions they differ more or less in the weight or efficiency which they assign to each. In a sum in multiplication you may gain the same product by using different factors; but if the product is to be constant, if you halve one factor, you must double another. Evolution is a product of many factors. One evolutionist lays more, another less, emphasis on natural selection, according as he assigns less or more efficiency to other forces or processes. Furthermore, evolutionists differ widely in questions of detail, and some of these subsidiary questions are of great practical importance and interest. It may be useful, therefore, to review these propositions in the light of the facts which we have gathered, and to see how they are interpreted, and what emphasis is laid on each by different thinkers.

The fundamental fact on which Mr. Darwin's theory rests is the " struggle for existence." Life is not something to be idly enjoyed, but a prize to be won; the world is not a play-ground, but an arena. And the severity of the struggle can scarcely be overrated. Only one or two of a host of runners reach the goal, the others die along the course. Concerning this there can be no doubt, and there is little room for difference of interpretation.

The struggle may take the form of a literal battle between two individuals, or of the individual with inclemency of climate or other destructive agents. More usually it is a competition, no more noticeable and no less real than that between merchants or manufacturers in the same line of trade.

The weeds in our gardens compete with the flowers for food, light, and place, and crowd them out unless prevented by man. And when the weeds alone remain, they crowd on each other until only a few of the hardiest and most vigorous survive. And flowers, by their nectar, color, and odor, compete for the visits of insects, which insure cross-fertilization. And fruits are frequently or usually the inducements by which plants compete for the aid of animals in the dissemination of their seeds. So there is everywhere competition and struggle; many fail and perish, few succeed and survive.

In a foot-race it is often very difficult to name the winner. Muscle alone does not win, not even good heart and lungs. Good judgment, patience, coolness, courage, many mental and moral qualities, are essential to the successful athlete. So in the struggle for life. The race is not always to the swift, nor the battle to the strong.

The total of " points " which wins this " grand

278 THE WHENCE AND THE WHITHER OF MAN

prize" is the aggregate of many items, some of which appear to us very insignificant. Hence, when we ask, "Who will survive?" the answer is necessarily vague. Mr. Darwin's answer is, Those best conformed to their environment; and Mr. Spencer's statement of the survival of the fittest means the same thing.

The judges who pronounce and execute the verdict of death, or award the prize of life, are the forces and conditions of environment. We have already considered the meaning of this word. Many of its forces and conditions are still unknown, or but very imperfectly understood. But known or unknown, visible or invisible, the result of their united action is the extinction or degradation of these individuals which deviate from certain fairly well-marked lines of development. We must keep clearly before our minds the fact that the world of living beings makes up by far the most important part of the environment of any individual plant or animal. Two plants may be equally well suited to the soil and climate of any region; but if one have a scanty development of root or leaf, or is for any reason more liable to attacks from insects or germs, other things being equal, it will in time be crowded out by its competitor. Worms are eaten by lower vertebrates, and these by higher. An animal's environment, like that of a merchant or manufacturer, is very largely a matter of the ability and methods of its competitors. And man, compelled to live in society, makes that part of the environment by which he is most largely moulded.

This process of extinction Mr. Darwin has called "natural selection." Natural selection is not a force, but a process, resulting from the combined action of the forces of environment. It is not a cause in any proper sense of the word, but a result of a myriad of interacting forces. The combination of these forces in a process of natural selection leading directly to a moral and spiritual goal demands an explanation in some ultimate cause. This explanation we have already tried to find.

It is a process of extinction. It favors the fittest, but only by leaving them to enjoy the food and place formerly claimed, or still furnished, by the less fit. In any advancing group, as the less fit are crowded out, and the better fitted gain more place and food and more rapid increase, the whole species becomes on an average better conformed. More abundant nourishment and increased vigor seem also to be accompanied by increased variation. And by the extinction of the less fit the probability is increased that more fit individuals will pair with one another and give rise to even fitter offspring, possessing perhaps new and still more valuable variations.

But if, of a group of weaker forms, those alone survive which adopt a parasitic life, those which in adult life move the least will survive and reproduce ; there will result the survival of the least muscular and nervous. This degeneration will continue until the species has sunken into equilibrium, so to speak, with its surroundings. Here natural selection works for degeneration. Sessile animals have had a similar history. But these parasitic and sessile forms had already been hopelessly distanced in the race for life. Their presence cannot impede the leaders ; indeed their survival is necessary to directly or indirectly furnish food for the better conformed. In the animal and plant world there is abundant room and advantage at the top.

Once more, natural selection works as a rule for the survival of individuals, only indirectly for that of organs composing, or of species including, these individuals. It may work for the development of a trait or structure which, while of no immediate advantage to the individual, increases the probability of its rearing a larger number of fitter offspring. Thus defence of the young by birds may be a disadvantage to the parent, but this is more than counterbalanced in the life of the species by the number of young coming to maturity and inheriting the trait. Even here natural selection favors the survival of the trait indirectly by sparing the descendants of the individual possessing it. Natural selection may always work on and through individuals without always working for their sole and selfish advantage.

In human society we find the selection of families, societies, nations, and civilizations going on, but mainly as the result of the survival of the fittest individuals.

There may very probably be a struggle for existence between organs or cells in the body of each individual. The amount of nutriment in the body is a more or less fixed quantity; and if one organ seizes more than its fair share, others may or must diminish for lack. But the limit to this usurpation must apparently be set by the crowding out of those individuals in which it is carried too far. Natural selection, so to speak, leaves the individual responsible for the distribution of the nutriment among the organs, and spares or destroys the individual as this usurpation proves for its advantage or disadvantage.

It makes its verdict much as the judges at a great poultry or dog show count the series of points, giving each one of them a certain value on a certain scale, and then award the prize to the individual having the highest aggregate on the whole series. Any such illustration is very liable to mislead ; I wish to emphasize that fitness to survive is determined by the aggregate of the qualities of an individual.

But an animal having one organ of great value or capacity may thus carry off the prize, even though its other organs deserve a much lower mark. This is the case with man. In almost every respect, except in brain and hand, he is surpassed by the carnivora, the cat, for example. But muscle may be marked, in making up the aggregate, on a scale of 500, and brain on a scale of 5,000, or perhaps of 50,000. A very slight difference in brain capacity outweighs a great superiority in muscle in the struggle between man and the carnivora, or between man and man.

The scale on which an organ is marked will be proportional to its usefulness under the conditions given at a given time. During the period of development of worms and lower vertebrates much muscle with a little brain was more useful than more brain with less muscle. Hence, as a rule, the more muscular survived ; the brain increasing slowly, at first apparently largely because of its correlation with muscle and sense-organs. At a later date muscle, tooth, and claw were more useful on the ground ; brain and hand in the trees. Hence carnivora ruled the ground, and certain arboreal apes became continually more anthropoid. At a later date brain became more useful even on the ground, and was marked on a higher scale, because it could invent traps and weapons against which muscle was of little avail. Just at present brain is of use to, and valued by, a large portion of society in proportion to its efficiency in making and selfishly spending money. But slowly and surely it is becoming of use as an organ of thought, for the sake of the truth which it can discover and incarnate.

Natural selection works thus apparently for the survival of the individuals possessing in the aggregate the most complete conformity to environment. Let us now imagine that an animal is so constructed as to be capable of variation along several disadvantageous or neutral lines, and along only one which is advantageous. The development would of course proceed along the advantageous line. Let us farther imagine that to the descendants of this individual two, and only two, advantageous lines of variations are allowed by its structure. Then natural selection would probably favor the decidedly advantageous line, if such there were. But as long as the structure of the animal allows variation along only a few lines, the two advantageous variations would, according to the law of probabilities, frequently occur in the same individual. The eggs and spermatozoa of two such individuals might not infrequently unite, and thus in time the two characteristics be inherited by a large fraction of the species.

And now let me quote from Mr. Spencer:

" But in proportion as the life grows complex—in proportion as a healthy existence cannot be secured by a large endowment of some one power, but demands many powers; in the same proportion do there arise obstacles to the increase of any particular power, by 'the preservation of favored races in the struggle for life.' As fast as the faculties are multiplied, so fast does it become possible for the several members of a species to have various kinds of superiorities over one another. While one saves its life by higher speed, another does the like by clearer vision, another by keener scent, another by quicker hearing, another by greater strength, another by unusual power of enduring cold or hunger, another by special sagacity, another by special timidity, another by special courage; and others by other bodily and mental attributes. Now it is unquestionably true that, other things equal, each of these attributes, giving its possessor an extra chance of life, is likely to be transmitted to posterity. But there seems no reason to suppose that it will be increased in subsequent generations by natural selection. That it may be thus increased, the individuals not possessing more than average endowments of it must be more frequently killed off than individuals highly endowed with it; and this can happen only when the attribute is one of greater importance, for the time being, than most of the other attributes. If those members of the species which have but ordinary shares of it, nevertheless survive by virtue of other superiorities which they severally possess, then it is not easy to see how this particular attribute can be developed by natural selection in subsequent generations. The probability seems rather to be that, by gamogenesis, this extra endowment will, on the average, be diminished in posterity—just serving in the long run to compensate the deficient endowments of other individuals whose special powers lie in other directions, and so to keep up the normal structure of the species. The working out of the process is here somewhat difficult to follow ; but it appears to me that as fast as the number of bodily and mental faculties increases, and as fast as the maintenance of life comes to depend less on the amount of any one, and more on the combined action of all, so fast does the production of specialties of character by natural selection alone become difficult. Particularly does this seem to be so with a species so multitudinous in its powers as mankind, and above all does it seem to be so with such of the human powers as have but minor shares in aiding the struggle for life—the æsthetic faculties for example."—Spencer, " Principles of Biology," § 166.

Can thus natural selection, acting upon fortuitous variations, be the sole guiding process concerned in progress? Must there not be some combining power to produce the higher individuals which are prerequisites to the working of natural selection?

We are considering the efficiency of natural selection in enhancing useful variations through a series of generations. Let us return to the distinction between productiveness and prospectiveness of social capital. Applied to variations productiveness means immediate advantage, prospectiveness the greater future and permanent returns. Now all persisting variations must, in animals below man, apparently be somewhat productive, else they would not continue, much less increase. Now the immediate return from prospective variations is often smaller than from productive. It looks at first as if productive variations would always be preserved by natural selection, and that prospective variations would not long advance. Yet in the muscular system variations valuable largely for their future value are neither few nor unimportant. How can the brain in its infancy develop until it gains supremacy over muscle, or muscle have done the same with digestion? Now a partial explanation of this is to be found in the correlation of organs. This is therefore a factor of vast importance in progress through evolution.

Progress in any one line demands correlated changes in many organs. Thus in the advance of annelids to insects the muscular system increases in relative bulk, and absolutely in complexity. But a change or increase in the muscle must be accompanied by corresponding changes in the motor-nerve fibrils; and these again would be useless unless accompanied by increased complexity and more or less readjustment of the cells and fibrils of the nerve-centres. And all these additions to, and readjustments of, the nerve-centres must take place without any disturbance of the other necessary adjustments already attained. This is no simple problem.

We will here neglect the fact that many other changes are going on simultaneously. Legs are being formed or moulded into jaws, the anterior segments are fusing into a head, and their ganglia into a brain ; an external skeleton is developing. Furthermore the increase of the muscular and nervous systems must be accompanied by increased powers of digestion, respiration, and excretion. Practically the whole body is being recast. We insist only on the necessity of simultaneous and parallel changes in muscles, nerves, and nerve-centres ; though what is true of these is true, in greater or less degree, of all the other organs.

You may answer that this is to be explained by the law of correlation of organs ; that when changes in one organ demand corresponding changes in another, these two change similarly and more or less at the same time and rate. But this is evidently not an explanation but a restatement of the fact. The question remains, What makes the organs vary simultaneously so as to always correspond to each other? The whole series of changes must to some extent be effected at once and in the same individual, if it is to be preserved by natural selection. Fortuitous variations here and there along the line of the series are of little or no avail. That the whole series of variations should happen to occur in one animal is altogether against the law of probabilities; if the favorable variation occurs in only a part of the series it remains useless until the corresponding variation has taken place in the other terms. And while the variation is thus awaiting its completion, so to speak, it is useless, and cannot be fostered by natural selection.

Evolution by means of fortuitous variations, combined and controlled only through natural selection, seems to me at least impossible; and this view is, I think, steadily gaining ground.

Natural selection, while a real and very important factor in evolution, cannot be its sole and exclusive explanation. It presupposes other factors, which we as yet but dimly perceive. And this does not impeach the validity of Mr. Darwin's theory any more than Newton's theory of gravitation is impeached by the fact that it offers no explanation as to why the apple falls or how bodies attract one another.

For natural selection explains the survival, but not the origin, of the fittest. Given a species or other group composed of more and less fit individuals and the fittest will survive. How does it come about that there are any more and less fit individuals? This brings us to the consideration of the subject of variation.

Let us begin with a simple case of change in the adult body. The workman grasps his tools day after day, and his hands become horny. The skin has evidently thickened, somewhat as on the soles of the feet. This is no mere mechanical result of pressure alone. Continuous pressure would produce the opposite result. But under the stimulus of intermittent pressure the capillaries, or smallest blood vessels, furnish more nutriment to the cells composing the lowest layer of the outer skin or epidermis. These cells, being better nourished, reproduce by division more rapidly, and the epidermis, becoming composed of a greater number of layers of cells, thickens. The outermost layers, being farthest from the blood supply, dry up and are packed together into a horny mass.

If I go out into the sunshine I become tanned. This again is not a direct and purely chemical or physical result of the sun's rays, but these have stimulated the cells of the skin to undergo certain modifications. Any change in the living body under changed conditions is not passive, but an active reaction to a stimulus furnished by the surroundings. The same stimulus may excite very different reactions in different individuals or species.

Early in this century a farmer, Seth Wright, found among his lambs a young ram with short legs and long body. The farmer kept the ram, reasoning that his short legs would prevent him from leading the flock over the farm-walls and fences. From this ram was descended the breed of ancon, or otter, sheep. Now the stimulus which had excited this variation must have been applied early' in embryonic life, or perhaps during the formation or maturing of the germ-cells themselves. Such a variation we call a congenital variation.

These cases are merely illustrations of the general truth that in every variation there are two factors concerned : the living being with its constitution and inherent tendencies and the external stimulus.

The courses of the different balls in a charge of grape-shot, hurled from a cannon, are evidently due to two sets of forces—1, their initial energy and the direction of their aim ; 2, the deflecting power of resisting objects or forces—or the different balls might roll with great velocity down a precipitous mountainside. In the first case velocity and direction of course would be determined largely by initial impulse ; in the second, by the attraction of the earth and by the inequalities of its surface.

In evolution, environment, roughly speaking, corresponds to these deflecting or attracting external objects or forces; inherent tendencies to initial impulse. If we lay great weight on initial tendencies, inherent in protoplasm from the very beginning, we shall probably lay less stress on natural selection as a guiding, directing process.

The great botanist, Nägeli, has propounded a most ingenious and elaborate theory of evolution, as dependent mainly on inherent initial tendency. We can notice only one or two of its salient points. All development is, according to his view, due to a tendency in the primitive living substance toward more complete division of labor and greater complexity. This tendency, which he calls progression, or the tendency toward perfection, is the result of the chemical and molecular structure of the formative controlling protoplasm (idioplasm) of the body, and is transmitted with other parental traits from generation to generation. And structural complexity thus increases like money at compound interest. Development is a process of unfolding or of realization of the possibilities of this tendency under the stimulus of surrounding influences. Environment plays an essential part in his system. But only such changes are transmissible to future generations as have resulted from modifications arising in the idioplasm. Descendants of plants which have varied under changed conditions revert, as a rule, to the old type, when returned to the old surroundings. And in the animal world effects of use and disuse are, according to his view, not transmissible.

Natural selection plays a very subordinate part. It is purely destructive. Given an infinity of place and nourishment—do away, that is, with all struggle and selection—and the living world would have advanced, purely by the force of the progressive tendency, just as far as it now has ; only there would have survived an indefinite number of intermediate forms. It would have differed from our present living world as the milky way does from the starry firmament.

He compares the plant kingdom to a great, luxurious tree, branching from its very base, whose twigs would represent the present stage of our different species. Left to itself it would put out a chaos of innumerable branches. Natural selection, like a gardener, prunes the tree into shape. Childen might imagine that the gardener caused the growth ; but the tree would have been broader and have branched more luxuriantly if left to itself.*

Every species must vary perpetually. Now this proposition is apparently not in accord with fact; for some have remained unchanged during immense periods. And natural selection, by removing the less fit, certainly appears to contribute to progress by raising the average of the species. The theory seems extreme and one-sided. And yet it has done great service by

* See Nägeli, " Theorie der Abstammungslehre," p. 18 ; also pp. 12, 118, 285.

19

calling in question the all-sufficiency of natural selection and the modifying power of environment, and by emphasizing, probably overmuch, the importance of initial inherent tendency, whose value has been entirely neglected by many evolutionists.

Lack of space compels us to leave unnoticed most of the exceedingly valuable suggestions of Nägeli's brilliant work.

It is still less possible to do any justice in a few words to Weismann's theory. Into its various modifications, as it has grown from year to year, we have no time to enter. And we must confine ourselves to his views of variation and heredity.

In studying protozoa we noticed that they reproduced by fission, each adult individual dividing into two young ones. There is therefore no old parent left to die. Natural death does not occur here, only death by violence or unfavorable conditions. The protozoa are immortal, not in the sense of the endless persistence of the individual, but of the absence of death. Heredity is here easily comprehensible, for one-half, or less frequently a smaller fraction, of the substance of the parent goes to form the new individual. There is direct continuity of substance from generation to generation.

But in volvox a change has taken place. The fertilized egg-cell, formed by the union of egg and spermatozoon, is a single cell, like the individual resulting from the conjugation or fusion of two protozoa. But in the many-celled individual, which develops out of the fertilized egg, there are two kinds of cells. 1. There are other egg-cells, like the first, each one of which can, under favorable conditions, develop into a multicellular individual like the parent. And the germ-cells (eggs and spermatozoa) of volvox are immortal like the protozoa. But, 2, there are nutritive, somatic cells, which nourish and transport the germ-cells, and after their discharge die. These somatic cells, being mortal, differ altogether from the germ-cells and the protozoa. The protoplasm must differ in chemical, or molecular, or other structure in the two cases, and we distinguish the germ-plasm of the germ-cells, resembling in certain respects Nägeli's idioplasm, from somatoplasm, which performs most of the functions of the cell. The somatoplasm arises from, and hence must be regarded as a modification of, the germ-plasm. The germ-plasm can increase indefinitely in the lapse of generations, increase of the somatoplasm is limited.

When a new individual develops, a certain portion of the germ-plasm of the egg is set aside and remains unchanged in structure. This, increasing in quantity, forms the reproductive elements for the next generation. The germ-plasm, which does not form the whole of each reproductive element, but only a part of the nucleus, is thus an exceedingly stable substance. And there is a just as real continuity of germ-plasm through successive generations of volvox, or of any higher plants or animals, as in successive generations of protozoa.

In certain plants there is an underground stem or rootstock, which grows perennially, and each year produces a plant from a bud at its end. This underground rootstock would represent the continuous germ-plasm of successive generations ; the plants which yearly arise from it would represent the successive generations of adult individuals, composed mainly of somatoplasm. Or we may imagine a long chain, with a pendant attached to each tenth or one-hundredth link. The links of the chain would represent the series of generations of germ-cells; the pendants, the adults of successive generations.

But any leaf of begonia can be made to develop into a new plant, giving rise to germ-cells. Here there must be scattered through the leaves of the plant small portions of germ-plasm, which generally remain dormant, and only under special conditions increase and give rise to germ-cells.

A large part of the germ-plasm of the fertilized egg is used to give rise to the somatoplasm composing the different systems of the embryo and adult. Weismann's explanation of this change of germ-plasm into somatoplasm is very ingenious, and depends upon his theory of the structure of the germ-plasm; and this latter theory forms the basis of his theory of evolution. It would take too long to state his theory of the structure of germ-plasm, but an illustration may present fairly clear all that is of special importance to us.

The molecules of germ-plasm are grouped in units, and these in an ascending series of units of continually increasing complexity, until at last we find the highest unit represented in the nucleus of the germ-cell. This grouping of molecules in units of increasing complexity is like the grouping of the men of an army in companies, regiments, brigades, divisions, etc.

To form the somatoplasm of the different tissues of the body, this complicated organization breaks up, as the egg divides, into an ever-increasing number of cells.

First, so to speak, the corps separate to preside over the formation of different body regions. Then the different divisions, brigades, and regiments, composing each next higher unit, separate, being detailed to form ever smaller portions of the body. The process of changing germ-plasm into somatoplasm is one of disintegration. The germ-plasm contains representatives of the whole army ; a somatic cell only representatives of one special arm of a special training. Germ-plasm in the egg is like Humpty-Dumpty on the wall ; somatoplasm, like Humpty-Dumpty after his great fall.

I use these rude illustrations to make clear one point: Germ-plasm can easily change into somatoplasm, but somatoplasm once formed can never be reconverted into germ-plasm, any more than the fallen hero of the nursery rhyme could ever be restored.

The germ-plasm is, according to Weismann, a very peculiar, complex, stable substance, continuous from generation to generation since the first appearance of life on the globe. It is in the body of the parent, but scarcely of it. Its relation to the body is like that of a plant to the soil or of a parasite to its host. It receives from the body practically only transport and⁁ nourishment. It is like a self-perpetuating, close corporation ; and the somatoplasm has no means of either controlling it or of gaining representation in it.

Says Weismann * : "The germ-cells are contained in the organism, and the external influences which affect them are intimately connected with the state of the organism in which they lie hid. If it be well nourished, the germ-cells will have abundant nutriment ; and, conversely, if it be weak and sickly, the germ-cells will

* Essays upon Heredity, p. 105.

be arrested in their growth. It is even possible that the effects of these influences may be more specialized; that is to say, they may act only upon certain parts of the germ-cells. But this is indeed very different from believing that the changes of the organism which result from external stimuli can be transmitted to the germ-cells and will redevelop in the next generation at the same time as that at which they arose in the parent, and in the same part of the organism."

But if the germ-plasm has this constitution and relation to the rest of the body, how is any variation possible? Different individuals of any species have slightly different congenital tendencies. Hence in the act of fertilization two germ-plasms of slightly different structure and tendency are mingled. The mingling of the two produces a germ-plasm and individual differing from both of the parents. Thus, according to Weismann's earlier view, the origin of variation was to be sought in sexual reproduction through the mingling of slightly different germ-plasms.

But how did these two germ-plasms come to be different? How was the variation started? To explain this Weismann went back to the unicellular protozoa. These animals are undoubtedly influenced by environment and vary under its stimuli. Here the variations were stamped upon the germ-plasm, and the commingling of these variously stamped germ-plasms has resulted in all the variations of higher animals.

Of late Weismann has modified and greatly improved this portion of his theory. He now accepts the view that external influences may act upon the germ-plasm not only in protozoa but also in all higher animals. Variation is thus due to the action or stimulus of external influences, supplemented by sexual reproduction.

But the very constitution of the germ-plasm and its relation to the body absolutely forbids the transmission of acquired somatic characteristics and of the special effects of use and disuse. Muscular activity promotes general health, and might thus conduce to better-nourished germ-cells and to more vigorous and therefore athletic descendants. The exercise of the muscles might possibly cause such a condition of the blood that the portion of the germ-plasm representing the muscular system of the next generation might be especially nourished or stimulated. Thus an athletic parent might produce more athletic children.

But let us imagine twin brothers of equal muscular development. One from childhood on exercises the lower half of his body; the other, the upper. Both take the same amount of exercise, and have perhaps equal muscular development, but located in different halves of the body. Now it is hard to conceive that it can make any difference in the nourishing or stimulating influence of the blood, whether the muscular activity resides in one half of the body or the other. The children might be exactly alike.

One man drives the pen, a second plays the piano, and a third wields a light hammer. All three use different muscles of the hand and arm. How can this use of special muscles stamp itself upon the germ-cells in such a way that the offspring will have these special muscles enlarged? Granting that external influences of environment and bodily condition may effect the germ-cells ; granting even that some of the most general effects of use and disuse might be transmitted, what warrant have we for believing that the special acquired characteristic can be transmitted? Weismann answers, None at all. The somatoplasm can only in the most general way affect the self-perpetuating, close corporation of the germ-plasm.*

There is thus, according to Weismann, nothing to direct variation to certain organs, or to guide and combine the variations of these organs along certain lines, except natural selection. To a certain extent variation may be limited by the very structure of the animal. But within these limits there are wide ranges where one variation is apparently just as likely to occur as another.

Within these wide limits variation appears to be fortuitous. Natural selection must wait until the individuals appear in which these variations occur already correlated, and then seize upon these individuals. It is apparently the only guiding, directing force. Linear variation, that is, a variation advancing continuously along one or very few straight lines, would appear to be impossible.

In Nägeli's theory initial tendency is overwhelmingly dominant; in Weismann's, natural selection is almighty.

Weismann's followers have received the name of Neo - Darwinians. The so - called Neo - Lamarckian school believes in the transmissibility of acquired characteristics, and of at least particular effects of use and disuse. The one theory is neither more nor less Darwinian than the other. For while Darwin emphasized natural selection, he accepted to a certain extent the transmission of special effects of use and disuse.

* Weismann, Essays, p. 286.

A special theory of heredity, pangenesis, has been accepted by many of the Neo-Lamarckian school. The theory of pangenesis, as propounded by Mr. Darwin, may be very briefly stated as follows : The cells in all parts of the body are continually throwing off germinal particles, or "gemmules." These become scattered through the body, grow, and multiply by division. On account of mutual attraction they unite in the reproductive glands to form eggs or spermatozoa. The germ-cells are thus the bearers of heredity because they contain samples, so to speak, of all the organs of the body.

In heredity, according to Weismann's theory, the egg is the centre of control, the continuous germ-plasm the source of all transmitted changes; according to Darwin's theory, the body is the source, and the egg is derived in great part at least from it. If you put to the two the time-honored question, Which is first, the owl or the egg? Weismann would announce, with emphasis, The egg ; Darwin would say, The owl. One proposition is the converse of the other, and most facts accord almost equally well with both theories.

In any family, devoted for generations to literary or artistic pursuits, the children show, as a rule, an aptitude for such pursuits not manifested by those of other families. According to the Neo-Lamarckian view, this inherited aptitude is to a certain extent the result of the constant exercise of these faculties through a series of generations. The active efforts and voluntary disposition of the parents have given an increased predisposition to the child. " Quite the reverse," says Weismann, "the increase of an organ in the course of generations does not depend upon the summation of exercise taken during single lives, but upon the summation of more favorable predispositions in the germ." " An organism cannot acquire anything unless it already possesses the predisposition to acquire it." *

We may accept or deny this last statement, but it is evident that facts like these, and indeed the origin of most or all characteristics involving use or disuse, may be explained almost equally well by either theory.

But as far as the transmission of effects of somatic changes is concerned, if protozoa undergo special modifications under the influence of external conditions, will not the germ-cells undergo special modification under the influence of changes in the somatoplasm which forms their immediate environment ? We must never forget the close relationship between all the cells of the body, and how slight a change in the body or its surroundings may conduce to sterility or fertility. Such isolation and independence in the body, on the part of the germ-cells, is opposed to all that we know of the organic unity of the body, whose cells have arisen by the differentiation of, and division of labor between, cells primitively alike. The facts of bud-variation, of changes in the parent stock due to grafting, and others, of which Mr. Darwin has given a summary in the eleventh chapter of the first volume of his " Plants and Animals under Domestication," have never been adequately explained by Weismann in accordance with his theory. He has perhaps succeeded in parrying their force by showing that some such explanation is conceivable ; they still point strongly against him.

* Weismann, Essays, pp. 85 and 171.

Wilson has good reason for his "steadily growing conviction that the cell is not a self-regulating mechanism in itself, that no cell is isolated, and that Weismann's fundamental proposition is false."

But, granting the force of these criticisms, the question still remains, Is the special effect of use or disuse transmissible? Would the blacksmith's son have a stronger right arm?

1. The isolation and independence of the germ-cells, which Weismann postulates as opposing this, can hardly be as great as he thinks. 2. It is in his view impossible to conceive how these acquired characteristics can in any way reach and affect the germ-cells in such a manner as to reappear in the next generation. 3. All variations can be explained by his own theory without such transmission. Why then believe that acquired characteristics can in some inconceivable way affect the germ-cells so as to reappear in the next generation, as long as all the facts can be explained in a more simple and easily conceivable manner?

As to his second argument, I would readily acknowledge that it is at present difficult or impossible for me to conceive how any cell can act upon another, except through the nutrient or other fluids which it can produce. But though I cannot conceive how one cell can affect another, I may be compelled to believe that it does so. And this Weismann readily acknowledges.

Driesch changed by pressure the relative position of the cells of a very young embryo, so that those which in a normal embryo would have produced one organ were now compelled, if used at all, to form quite a different one. And yet these displaced cells formed the organ required of cells normally occupying this

new position, not the one for which they were normally
intended. And the organ which they would have
builded in a normal embryo was now formed by other
cells transferred to their rightful place.

What made them thus change? Not change of substance or structure, for the slight pressure could hardly
have modified this. Not change of nutriment. The
only visible or easily conceivable change was in position relative to other cells of the embryo.

Let us in imagination simplify Driesch's experiment,
for the sake of gaining a clearer view of its meaning.
In a certain embryo at an early stage are certain cells
whose descendants should form the lining of the intestine and be used in the adult for digestion. A
second set of cells should form muscle endowed mainly
with contractility. When these two sets of cells, or
some of them, exchange positions in the embryo, they
exchange lines of development. The first set now
form muscle, the second digestive tissue. The only
change has been in their relative positions. Driesch
maintains, therefore, that the goal of development in
any embryonic cell is determined not by structure or
nutriment but by position. And this would seem to
be true of the cells of the earliest embryonic stages.

Certain other experiments point in the same direction. Cut a hydra into equal halves and each half
will form a complete animal. The lower half forms a
new top, with mouth and tentacles ; the upper half,
a new base. Cut the other hydra a hair's-breadth
farther up. The same layer of cells which in the
first animal formed the lower exposed surface of the
upper half now forms the upper exposed surface of
the lower half. And with this change of position it

has changed its line of development; it will now give rise to a new upper half, not a base as before. The same experiment can be tried on certain worms with similar results, only head and tail differ far more than top and base of hydra. Difference in the position of cells has made vast difference in their line of development. Now in both embryo and adult there must be some directing influence guiding these cells. What is it?

An army is more than a mob of individuals; it is individuals plus organization, discipline, authority. A republic is not square miles of territory and thousands or millions of inhabitants. It is these plus organization, central government. Webster claimed that the central government was, and had to be, before the states. The organism cannot exist without its parts; it has a very real existence in and through them. It can coerce them. The state may be an abstraction, but it is one against which it is usually fatal to rebel, and which can say to a citizen, Go and be hanged, and he straightway mounts the scaffold. Now these are analogies and prove nothing. But in so far as they throw light on the essential idea of an organism, they may aid us in gaining a right view of our "cell republic."

Says Whitman in a very interesting article on the "Inadequacy of the Cell-Theory": "That organization precedes cell-formation and regulates it, rather than the reverse, is a conclusion that forces itself upon us from many sides." "The structure which we see in a cell-mosaic is something superadded to organization, not itself the foundation of organization. Comparative embryology reminds us at every turn that the organism dominates cell-formation, using for the same purpose one, several, or many cells, massing its material and directing its movements, and shaping its organs as if cells did not exist, or as if they existed only in complete subordination to its will, if I may so speak. The organization of the egg is carried forward to the adult as an unbroken physiological unity, or individuality, through all modifications and transformations." And Wilson, Whitman, Hertwig, and others urge " that the organism as a whole controls the formative processes going on in each part " of the embryo. And many years ago Huxley wrote, " They (the cells) are no more the producers of the vital phenomena than the shells scattered along the sea-beach are the instruments by which the gravitative force of the moon acts upon the ocean. Like these, the cells mark only where the vital tides have been, and how they have acted." *

"Interaction of cells " can help us but little. For how can neighboring cells direct others placed in a new position? The expression, if not positively misleading and untrue, is at the best only a restatement of fact. It certainly offers no explanation. Floodtide is not due to the interaction of particles of water, though this may influence the form of the waves.

The centre of control is therefore not to be sought in individual cells, whether germ-cells or somatic, but in the organism. And it is the whole organism, one and indivisible, which controls in germ, embryo, and adult, in egg and owl. This individuality, or whatever you will call it, impresses itself upon de-

* See articles by Whitman and Wilson, Journal of Morphology, vol. viii., pp. 649, 607, etc.

veloping somatic cells, moulding them into appropriate organs, and upon germ-cells in process of formation, moulding them so that they may continue its sway. The muscle, modified by use or disuse, is a better expression of the individuality of its possessor, and the same individuality moulds similarly and simultaneously the germ-cells. Both are different expressions or manifestations of the same individuality. Only slowly does the individuality mould the muscles and nerves of the adult body to its use. Still more slow may be the moulding of the still more refractory germ-plasm, if such there be. But the moulding process goes on parallel in the two cases.

But Weismann's argument rests not merely upon any difficulty or impossibility of the transmissibility of acquired characteristics. His argument is rather that all facts can be better explained by his theory without postulating or accepting such transmission, cases of which have never been absolutely proven. But the question is not whether his theory offers a possible explanation of the facts, but whether it is the most probable explanation of all the facts. No one would deny, I think, that the continuity of the germ-plasm offers the best and most natural explanation of heredity; and that variations could be produced by the influence on the germ-plasm of external conditions seems entirely probable.

But when we consider the aggregation of these variations in a process of evolution, his theory seems unsatisfactory. We have already seen that what we commonly call a variation involves not one change, but a series of changes, each term of which is necessary. Muscle, nerve, and ganglion must all vary simultaneously and correspondingly. Correlation and combination are just as essential as variation. And evolution often demands the disappearance of less fit structures just as much as the advance of the fittest. Says Osborne, "It is misleading to base our theory of evolution and heredity solely upon entire organs; in the hand and foot we have numerous cases of muscles in close contiguity, one steadily developing, the other degenerating." Weismann offers the explanation that "if the average amount of food which an animal can assimilate every day remains constant for a considerable time, it follows that a strong influx toward one organ must be accompanied by a drain upon others, and this tendency will increase, from generation to generation, in proportion to the development of the growing organ, which is favored by natural selection in its increased blood-supply, etc.; while the operation of natural selection has also determined the organ which can bear a corresponding loss without detriment to the organism as a whole." *

Here again natural selection of individuals, not the diminished supply of nutriment, has to determine which of many muscles shall be poorly fed and which favored. But natural selection can favor special organs only indirectly through the individuals which possess such organs. Variation is fortuitous, and there is nothing, except natural selection, to combine or direct them. And, I think, we have already seen that any theory which neglects or excludes such directing and combining agencies must be unsatisfactory and inadequate. Weismann has promised us an explanation of correlation of variation in accordance with his

* Weismann, Essays, p. 88.

theory; and if such an explanation can be made, it would remove one of the strongest objections. But for the present the objection has very great weight.

Furthermore, as Osborne has insisted, linear variations, or variations proceeding along certain single and well-marked lines, would seem inexplicable by, if not fatal to, Weismann's theory. And yet Osborne, Cope, and others have shown that the teeth of mammals have developed steadily along well - marked lines. They have apparently not resulted at all by selection from a host of fortuitous variations.

Says Osborne in his "Cartwright Lectures"*: "It is evident that use and disuse characterize all the centres of evolution; that changes of structure are slowly following on changes of function or habit. In eight independent regions of evolution in the human body there are upward of twenty developing organs, upward of thirty degenerating organs." Now this parallelism, through a long series of generations, between the evolution of organs, their advance or degeneration, and the use or disuse of these same organs, that is, of the habits of the individual, is certainly of great significance. It must have an explanation; and the most natural one would seem to be the transmission of the effects of use and disuse.

On the whole Osborne's verdict would seem just: The Neo-Lamarckian theory fails to explain heredity, Weismann's theory does not explain evolution. But, if the effects of use and disuse are transmitted, correlation of variation is to be expected. Muscle, nerve, and ganglion all vary in correlation because they are all used together and in like degree. Evolution and

* American Naturalist, vols. xxv. and xxvi.

20

degeneration of muscles in hand and foot go on side by side, because some are used and some are disused. Centres of use and disuse must be centres of evolution. And there would be as many distinct centres of evolution in different parts of the body as there were centres of use and disuse. And between these centres there might be no correlation except that of use and disuse. Brain, muscles, and jaws would develop simultaneously in the ancestors of insects. And the effects of use and disuse, transmitted through a series of generations, would be cumulative. The species advances rapidly because all its members have in general the same habits; the same parts are advancing or degenerating, although at different rates, in all its individuals. An animal having an organ highly developed is far less likely to pair with one having a lower development of the same organ. The Neo-Lamarckian theory supplies thus what is lacking in the Neo-Darwinian.

In lower forms, like hydra, of simple structure and comparatively few possibilites of variation, natural selection is dominant. In higher forms, like vertebrates, and especially in man, it is of decidedly subordinate value as a promoter of evolution. For man, as we have seen, is a marvellously complex being. The great difficulty in his case is not so much to quickly gain new and favorable variations as to keep all the organs and powers of the body steadily advancing side by side. Natural selection has in man the important but subordinate position of the judge in a criminal court, to pronounce the death verdict on the hopeless and incorrigible.

Both Neo-Darwinians and Neo-Lamarckians have

erred in being too exclusively mechanical in their
theories. It is the main business of the scientific man
to discover and study mechanisms. But he must remember that mechanism does not produce force, it
only transmits it. If he maintains that he has nothing to do with anything outside of mechanism, that
the invisible and imponderable force lies outside of
his domain, he has handed over to metaphysics the
fairest and richest portion of his realm. In our fear
of being metaphysical we have swung to another extreme, and have lost sight of valuable truth which lay
at the bottom of the old vitalistic theories. Cells,
tissues, and organs are but channels along which the
flood of life-force flows. Boveri has well said, "There
is too much intelligence (Verstand) in nature for any
purely mechanical theory to be possible."

Each theory contains important truth. Nägeli's
view of the importance of initial tendencies, inherent
in the original living substance, is too often under-
valued. My own conviction, at least, is steadily
strengthening that, without some such original tendency or aim, evolution would never have reached its
present culmination in man. His error lies in emphasizing this factor too exclusively. The fundamental proposition of Weismann's theory, that heredity is due to continuity of germ-plasm, seems to contain
important truth. But we need not therefore accept
his theory of a germ-plasm so isolated and independent as to be beyond control or influence by the habits
of the body. The importance of use and disuse, and
the transmissibility of their effects, would seem to
supply a factor essential to evolution. Weismann has
done good service in emphasizing the stability of the

germ-plasm. Evolution is always slow, and, for that very reason, sure.

If these conclusions are correct, they have an important practical bearing. Struggle and effort are essential to progress. Not inborn talent alone, but the use which one makes of it, counts in evolution. The effects of use and disuse are cumulative. The hard-fought battle of past generations becomes an easy victory in the present, just because of the strength acquired and handed down from the past struggle. Persistent variation toward evil is in time weeded out by natural selection. And, while evil remains in the world, we are to lay up stores of strength for ourselves and our descendants by sturdily fighting it. But the effects of right living through a hundred generations are not overcome by the criminal life of one or two. Evil surroundings weigh more in producing criminals than heredity, and their children are not irreclaimable.

The struggles and victories of each one of us encourage the rest. There is, to borrow Mr. Huxley's language, not only a survival of the fittest, but a fitting of as many as possible to survive. And in the midst of the hardest struggle there is the peace which comes from the assurance of a glorious triumph.

Condensed Chart of Development of the Main Line of the Animal Kingdom leading to Man.

Phylogenetic Series.	New Attainments.	Organs Approaching Culmination.	Most Rapidly Advancing Organs.	Dominant Function.	Dominant Mental (or Nervous) Action.	Sequence of Perceptions.	Sequence of Motives.	Environment Marks For.
Amoeba.	Cell.					Touch. Smell.	Hunger.	
Volvox.	Somatic and reproductive cells.		Reproductive.	Reproduction.		Touch. Smell.	Hunger.	
Hydra.	Simple reproductive organs. Gastro vascular cavity. (Tissues).		Reproductive.	Reproduction.	Reflex.	Touch. Smell.	Hunger.	Rapid reproduction and good digestion.
Turbellaria.	Complex reproductive organs. Supra-oca. Ganglion and cords. Sense organs. Body Wall.	Reproductive.	Digestive.	Reproduction.	Reflex.	Touch. Smell.	Hunger.	
Amulid.	Perivisceral Cavity. Intestine. Circulatory system. Nephridia. Visual eyes.			Digestion Muscular.	Reflex.	Touch. Smell.	Hunger.	
Primitive Vertebrate.	Notochord. Fins.			Digestion Muscular.	Instinct.	?		
Fish.	Backbone (incomplete). Paired Fins. Jaws from Branchial Arches. Simple heart. Air bladder. Brain.	Digestive.	Muscles.	Digestion Muscular.	Instinct.	Hearing. Sight.		Strength and activity.
Amphibian.	Legs. Lungs. Cerebrum increases from this form on.		Muscles.	Digestion Muscular.	Instinct.	Hearing. Sight.	Fear and other prudential considerations.	
Reptile.	Double heart.		Muscles and appendages.	Muscular.	Instinct ?	Hearing. Sight.		
Lower Placental Mammals.	Constant high temperature. Placenta.	Muscle.	Muscles and appendages.	Muscular.	Instinct ? ?	Hearing. Sight.		
Ape.	Erect posture. Hand. Large cerebrum.		Brain.	Muscular. Nervous.	Intelligence.	Mental perception. Understanding. Association.	"	" ? (Shrewdness?)
Man.	Very large cerebrum. Personality.		Brain.	Mind.*	Intelligence.	Reason.*	Love of man. Truth. Right.	Shrewdness. Righteousness* and unselfishness.*

* Apparently capable of indefinite development.

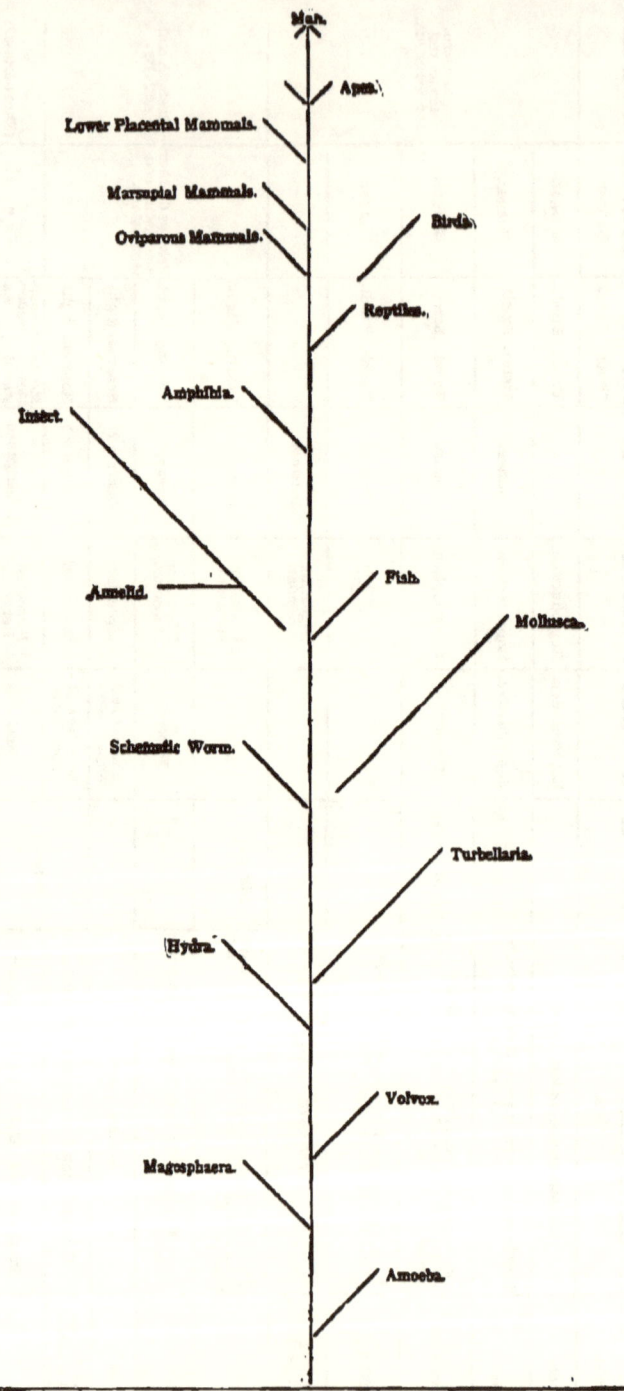

PHYLOGENETIC CHART OF PRINCIPAL TYPES OF
ANIMAL LIFE.

INDEX

The Ely Lectures for 1897

THE BIBLE AND ISLAM: OR,

THE INFLUENCE OF THE OLD AND NEW TESTAMENTS ON THE RELIGION OF MOHAMMED

By HENRY PRESERVED SMITH, D.D.

12mo, 319 pp., $1.50

CONTENTS

I. The Apostle of Allah.
II. The Common Basis in Heathenism.
III. The Koran Narratives.
IV. The Doctrine of God.
V. The Divine Government.

VI. Revelation and Prophecy.
VII. Sin and Salvation.
VIII. The Service of God.
IX. The Future Life.
X. Church and State.

"We should be inclined to regard this volume as perhaps the very best for one who desired to get a clear understanding of the doctrines rather than of the practical workings of Mohammedanism."
—*The Outlook.*

"The general reader will not meet with a more complete compendium of the religious teachings of the Prophet of Arabia."
—New York *Commercial Advertiser.*

The Ely Lectures for 1891

ORIENTAL RELIGIONS AND CHRISTIANITY

A COURSE OF LECTURES DELIVERED BEFORE THE STUDENTS OF UNION THEOLOGICAL SEMINARY, NEW YORK

By FRANK F. ELLINWOOD, D.D.

Secretary of the Presbyterian Board of Foreign Missions.

12mo, 384 pp., $1.75

CONTENTS

I. The Need of Understanding the False Religions.
II. The Methods of the Early Christian Church in Dealing with Heathenism.
III. The Successive Developments of Hinduism.
IV. The Bhagavad Gita and the New Testament.
V. Buddhism and Christianity.
VI. Mohammedanism Past and Present.
VII. The Traces of a Primitive Monotheism.
VIII. Indirect Tributes of Heathen Systems to the Doctrines of the Bible.
IX. Ethical Tendencies of the Eastern and the Western Philosophies.
X. The Divine Supremacy of the Christian Faith.

" The special value of this volume is in its careful differentiation of the schools of religionists in the East, and the distinct points of antagonism of the very fundamental ideas of Oriental religions toward the religion of Jesus."—*Outlook*.

" A more instructive book has not been issued for years."
—New York *Observer*.

" The author has read widely, reflected carefully, and written ably."
—*Congregationalist*.

" It is a book which we can most heartily commend to every pastor and to every intelligent student, of the work which the Church is called to do in the world."—*The Missionary*.

The Ely Lectures for 1890

THE EVIDENCE OF CHRISTIAN EXPERIENCE

By LEWIS FRENCH STEARNS, D.D.

12mo, 473 pp., $2.00

CONTENTS

I. The Evidences of To-day.
II. Philosophical Presuppositions—Theistic.
III. Philosophical Presuppositions—Anthropological.
IV. The Genesis of the Evidence.
V. The Growth of the Evidence.
VI. The Verification of the Evidence.
VII. Philosophical Objections.
VIII. Theological Objections.
IX. Relation to other Evidences.
X. Relation to other Evidences—Conclusion.

" His presentation of the certainty, reality, and scientific character of the facts in a Christian consciousness is very strong."—*The Lutheran*.

" An important contribution to the library of apologetics."
—*Living Church*.

" A good and useful work."—*The Churchman*.

" The tone and spirit which pervade them are worthy of the theme, and the style is excellent. There is nothing of either cant or pedantry in the treatment. There is simplicity, directness, and freshness of manner which strongly win and hold the reader."—Chicago *Advance*.

CHARLES SCRIBNER'S SONS

153-157 FIFTH AVENUE, NEW YORK

www.ingramcontent.com/pod-product-compliance
Lightning Source LLC
Chambersburg PA
CBHW020943030726
47496CB00005B/1327